Causal Inference in Pharmaceutical Statistics

Causal Inference in Pharmaceutical Statistics introduces the basic concepts and fundamental methods of causal inference relevant to pharmaceutical statistics. This book covers causal thinking for different types of commonly used study designs in the pharmaceutical industry, including but not limited to randomized controlled clinical trials, longitudinal studies, single-arm clinical trials with external controls, and real-world evidence studies. The book starts with the central questions in drug development and licensing, takes the reader through the basic concepts and methods via different study types and through different stages, and concludes with a roadmap to conduct causal inference in clinical studies. The book is intended for clinical statisticians and epidemiologists working in the pharmaceutical industry. It will also be useful to graduate students in statistics, biostatistics, and data science looking to pursue a career in the pharmaceutical industry.

Key Features:

- Causal inference book for clinical statisticians in the pharmaceutical industry
- Introductory level on the most important concepts and methods
- Align with FDA and ICH guidance documents
- Across different stages of clinical studies
- Cover a variety of commonly used study designs

Yixin Fang, Ph.D. is Director of Statistics and Research Fellow at AbbVie Inc. He obtained his Ph.D. in Statistics from Columbia University and is an experienced statistician and data scientist who has a history of working in both the biopharmaceutical industry and academia.

Chapman & Hall/CRC Biostatistics Series

Series Editors
Shein-Chung Chow, Duke University School of Medicine, USA
Byron Jones, Novartis Pharma AG, Switzerland
Jen-pei Liu, National Taiwan University, Taiwan
Karl E. Peace, Georgia Southern University, USA
Bruce W. Turnbull, Cornell University, USA

Recently Published Titles
Design and Analysis of Pragmatic Trials
Song Zhang, Chul Ahn and Hong Zhu

ROC Analysis for Classification and Prediction in Practice
Christos Nakas, Leonidas Bantis and Constantine Gatsonis

Controlled Epidemiological Studies
Marie Reilly

Statistical Methods in Health Disparity Research
J. Sunil Rao, Ph.D.

Case Studies in Innovative Clinical Trials
Edited by Binbing Yu and Kristine Broglio

Value of Information for Healthcare Decision Making
Edited by Anna Heath, Natalia Kunst, and Christopher Jackson

Probability Modeling and Statistical Inference in Cancer Screening
Dongfeng Wu

Development of Gene Therapies: Strategic, Scientific, Regulatory, and Access Considerations
Edited by Avery McIntosh and Oleksandr Sverdlov

Bayesian Precision Medicine
Peter F. Thall

Statistical Methods Dynamic Disease Screening Spatio-Temporal Disease Surveillance
Peihua Qiu

Causal Inference in Pharmaceutical Statistics
Yixin Fang

For more information about this series, please visit: https://www.routledge.com/Chapman--Hall-CRC-Biostatistics-Series/book-series/CHBIOSTATIS

Causal Inference in Pharmaceutical Statistics

Yixin Fang

CRC Press is an imprint of the
Taylor & Francis Group, an **informa** business
A CHAPMAN & HALL BOOK

Designed cover image: © Shutterstock Stock Illustration: 197106719, Illustration Contributor Mopic

First edition published 2024
by CRC Press
2385 NW Executive Center Drive, Suite 320, Boca Raton FL 33431

and by CRC Press
4 Park Square, Milton Park, Abingdon, Oxon, OX14 4RN

CRC Press is an imprint of Taylor & Francis Group, LLC

ISBN: 978-1-032-56014-4 (hbk)
ISBN: 978-1-032-56015-1 (pbk)
ISBN: 978-1-003-43337-8 (ebk)

DOI: 10.1201/9781003433378

Typeset in CMR10
by KnowledgeWorks Global Ltd.

Publisher's note: This book has been prepared from camera-ready copy provided by the authors.

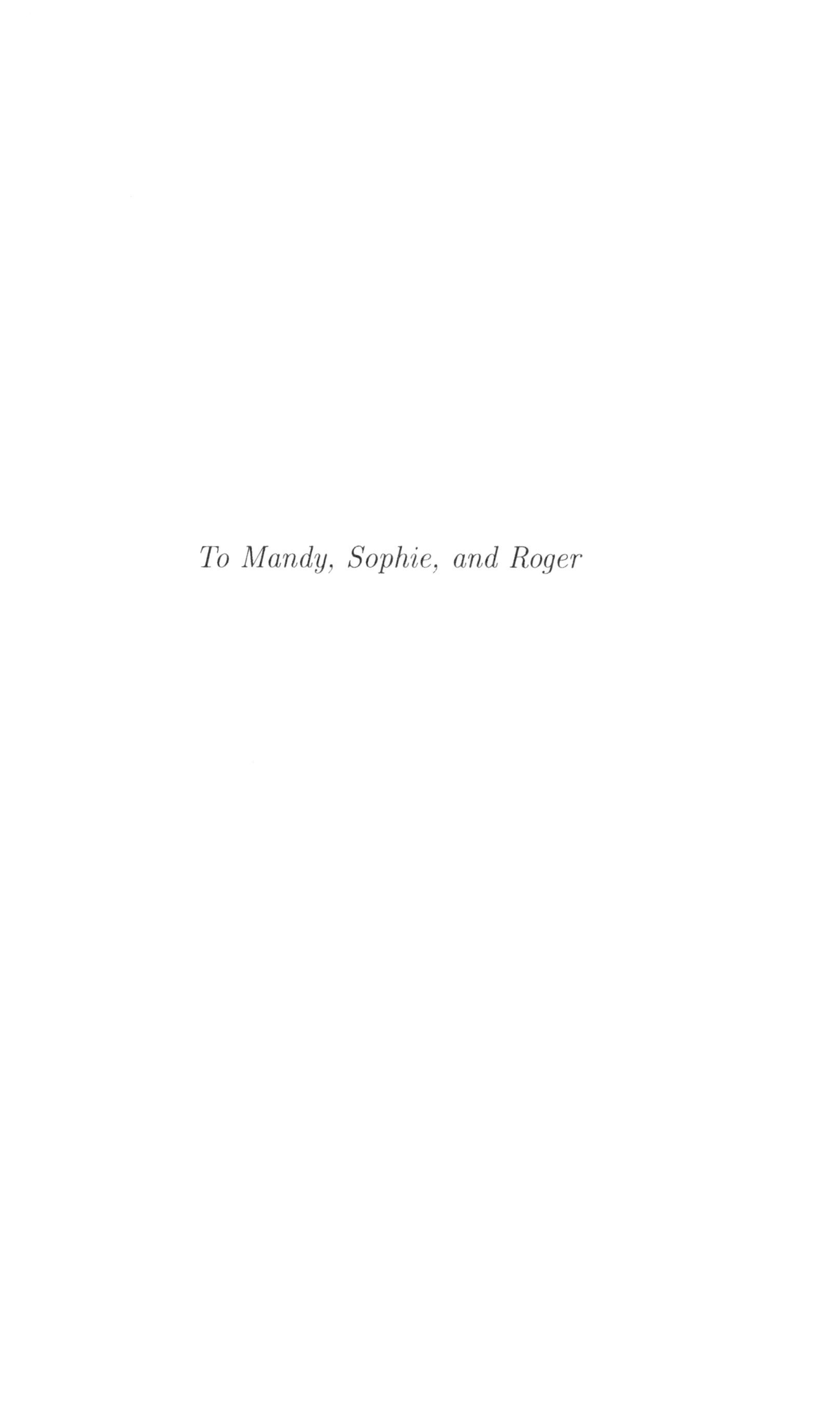

To Mandy, Sophie, and Roger

Contents

Preface **xiii**

1 Introduction **1**
1.1 Central Questions . . . 1
1.2 Potential Outcomes . . . 3
1.3 Estimand . . . 5
1.3.1 The PROTECT checklist . . . 5
1.3.2 Estimand for a given population . . . 6
1.3.3 Estimand for a given super-population . . . 8
1.3.4 Internal validity and external validity . . . 8
1.4 Probability and Statistics . . . 10
1.4.1 Probability . . . 10
1.4.2 Directed acyclic graphs . . . 13
1.4.3 Statistics . . . 15
1.5 Exercises . . . 16

2 Randomized Controlled Clinical Trials **19**
2.1 Randomization and Blinding . . . 19
2.2 Estimand . . . 21
2.2.1 Causal estimand . . . 21
2.2.2 Statistical estimand . . . 22
2.3 Estimator . . . 24
2.3.1 Expectation of the estimator . . . 24
2.3.2 Variance of the estimator . . . 26
2.3.3 Statistical inference . . . 29
2.4 Common Types of Randomization . . . 30
2.4.1 Simple randomization . . . 30
2.4.2 Block randomization . . . 33
2.4.3 Stratified randomization . . . 34
2.5 Exercises . . . 38

3 Missing Data Handling **40**
3.1 Missing Data . . . 40
3.2 Intent-to-treat Effect . . . 41
3.2.1 Scenario one . . . 41
3.2.2 Scenario two . . . 42

3.3 Per-protocol Effect 45
3.4 Sources of Missing Data 48
3.4.1 Intercurrent events 48
3.4.2 Missing data that are consequences of ICEs 49
3.4.3 Missing data that are not consequences of ICEs 50
3.5 Appendix 51
3.6 Exercises 52

4 Intercurrent Events Handling 54
4.1 Five Strategies 54
4.1.1 The treatment policy strategy 55
4.1.2 The hypothetical strategy 57
4.1.3 The composite variable strategy 59
4.1.4 The while on treatment strategy 62
4.1.5 The principal stratum strategy 63
4.2 Combinations of Strategies 70
4.3 Time-to-event Outcome 72
4.3.1 Censoring 72
4.3.2 The treatment policy strategy 73
4.3.3 The hypothetical strategy 73
4.3.4 The composite variable strategy 73
4.3.5 The while on treatment strategy 73
4.3.6 The principal stratum strategy 74
4.3.7 The competing risk strategy 74
4.4 Sample Size Calculation 74
4.4.1 The treatment policy strategy 75
4.4.2 The hypothetical strategy 76
4.4.3 The composite variable strategy 76
4.4.4 The while on treatment strategy 76
4.4.5 The principal stratum strategy 77
4.5 Exercises 77

5 Longitudinal Studies 79
5.1 Continuous or Binary Outcome 79
5.1.1 The intent-to-treat effect 80
5.1.2 The per-protocol effect 86
5.2 Time-to-event Outcome 89
5.2.1 The intent-to-treat effect 91
5.2.2 The per-protocol effect 92
5.3 Treatment Regimes 92
5.3.1 Dynamic treatment regimes 94
5.3.2 SMART design 95
5.4 Exercises 97

6 Real-World Evidence Studies 99
6.1 RWE Studies . . . 99
6.1.1 Pragmatic RCTs . . . 99
6.1.2 Observational studies . . . 100
6.1.3 Externally controlled trials . . . 101
6.2 Confounding Bias . . . 101
6.2.1 No unmeasured confounder . . . 102
6.2.2 Unmeasured confounders . . . 105
6.2.3 Proxy variables . . . 107
6.3 Longitudinal Cohort Studies . . . 109
6.3.1 Causal estimand . . . 109
6.3.2 Identifiability assumptions . . . 110
6.3.3 Identification . . . 111
6.4 Externally Controlled Trials . . . 113
6.4.1 Causal estimand . . . 113
6.4.2 Identification . . . 114
6.5 Appendix . . . 115
6.6 Exercises . . . 117

7 The Art of Estimation (I): M-estimation 119
7.1 Introduction . . . 119
7.2 M-estimation . . . 120
7.2.1 M-estimator . . . 120
7.2.2 Asymptotic linearity . . . 122
7.2.3 Regularity . . . 122
7.3 G-computation Estimator . . . 127
7.3.1 Plug-in estimator . . . 127
7.3.2 MLE . . . 128
7.3.3 Asymptotic variance . . . 130
7.3.4 Influence function . . . 131
7.4 Inverse Probability Weighted Estimator . . . 131
7.4.1 IPW estimator . . . 131
7.4.2 Asymptotic variance . . . 132
7.4.3 Influence function . . . 134
7.5 Augmented Inverse Probability Weighted Estimator . . . 134
7.5.1 A class of estimators . . . 134
7.5.2 Asymptotic variances . . . 135
7.5.3 AIPW estimator . . . 136
7.5.4 Double robustness . . . 138
7.5.5 Influence function . . . 139
7.6 Exercises . . . 140

8 The Art of Estimation (II): TMLE **142**
8.1 Semiparametric Statistics 142
8.1.1 Semiparametric estimators 142
8.1.2 Super learner 143
8.1.3 Semiparametric estimators based on super learner . . . 144
8.2 Asymptotic Variances of Semiparametric Estimators 145
8.2.1 Parametric submodels 145
8.2.2 The fundamental theorem of regularity 147
8.2.3 Influence function of MLE-SL estimator 148
8.2.4 Influence function of IPW-SL estimator 150
8.2.5 Double robustness of AIPW-SL estimator 152
8.2.6 Efficient influence function 153
8.3 The Targeted Learning Framework 154
8.3.1 Mini-roadmap 155
8.3.2 TMLE 155
8.3.3 Double robustness 156
8.4 A Shortcut to Derive Efficient Influence Functions 157
8.4.1 The efficient influence function for ATE 157
8.4.2 The efficient influence function for ATT 159
8.4.3 Missing data due to analysis dropout 160
8.5 Discussion 162
8.5.1 How to select covariates? 162
8.5.2 How to handle missing covariates? 163
8.5.3 How to use TMLE for RCTs? 163
8.5.4 How to implement TMLE? 164
8.6 Exercises 165

9 The Art of Estimation (III): LTMLE **167**
9.1 Longitudinal Cohort Studies 167
9.1.1 Causal estimand 167
9.1.2 Identification 168
9.1.3 Efficient influence function 171
9.1.4 LTMLE 172
9.1.5 ATE estimand 173
9.2 Missing Data 174
9.2.1 Monotone missing 174
9.2.2 Non-monotone missing 177
9.3 Implementation 177
9.4 Exercises 179

10 Sensitivity Analysis **180**
10.1 Introduction 180
10.2 Sensitivity Analysis for Identifiability Assumptions 182
10.2.1 The consistency assumption 182
10.2.2 The exchangeability assumption 183

10.2.3 The positivity assumption 185
10.3 Sensitivity Analysis for the MAR Assumption 186
10.3.1 A class of reference-based imputation models 186
10.3.2 Sequential modeling 188
10.4 Appendix 188
10.5 Exercises 190

11 A Roadmap for Causal Inference 191
11.1 Introduction 191
11.2 Roadmap 192
11.2.1 Study protocol 192
11.2.2 Data collection 194
11.2.3 Statistical analysis plan 194
11.2.4 Clinical study report 195
11.3 A Plasmode Case Study 195
11.3.1 Research question 196
11.3.2 Study design 196
11.3.3 Causal estimand 197
11.3.4 Data 198
11.3.5 Statistical estimand 199
11.3.6 Estimator 199
11.3.7 Estimate 199
11.3.8 Sensitivity analysis 200
11.3.9 Evidence 201
11.4 Exercises 202

12 Applications of the Roadmap 203
12.1 Introduction 203
12.2 Applications to RCTs 203
12.2.1 RCTs with a single follow-up 204
12.2.2 Longitudinal RCTs 207
12.2.3 RCTs with time-to-event outcome 210
12.3 Applications to Cohort Studies 212
12.3.1 Cohort studies with a single follow-up 213
12.3.2 Externally controlled trials 215
12.3.3 Longitudinal cohort studies 218
12.4 Exercises 221

Bibliography 223

Index 229

Preface

My journey of learning causal inference started in the summer of 2005, when I was a graduate student in the Department of Statistics at Columbia University. In that year, the department invited two renowned professors from Harvard University—Professor Donald Rubin and Professor James Robins—to teach summer courses on causal inference, respectively. Eighteen years later, I still feel the luck and excitement that I was one of the students sitting in their classes and listening to their thought-provoking lectures. After the two summer courses, I started reading *Causality*, a book by Professor Judea Pearl, now and then. And, fortunately, when I was one of the members of the Partnership of Public Health at Georgia State University, I participated in organizing a keynote speech given by Professor Pearl on causality.

My ongoing trek of applying causal inference started in 2011, when I joined the faculty of the Division of Biostatistics at New York University School of Medicine, serving as lead statistician for several pragmatic randomized clinical trials and observational studies. I started to encounter many real-life problems to which I could always find solutions from the online draft of *Causal Inference: What If*, a book by Professors Miguel Hernan and James Robins, of which I have been waiting for many years to purchase a hard copy.

Since I joined the Department of Data and Statistical Sciences at AbbVie in 2019, I have often heard statements such as "because we are doing randomized controlled trials, we don't need to know causal inference." I begin to wonder why there are such false statements. As a statistician who came to the pharmaceutical industry from many years of academic career, I believe it is necessary to add some training in causal inference to the resumes of our hard-working clinical statisticians. And this is the motivation for my starting this book project.

Eventually this book would never be comparable to any of the existing monographs in the field of causal inferences including but not limited to:

- *Causality* by Pearl (2000, 2009; two editions)
- *Causal Inference* by Imbens and Rubin (2015)
- *Causal Inference in Statistics* by Pearl, Glymour, and Jewell (2016)
- *Targeted Learning* by van der Laan and Rose (2011, 2018; two books)
- *The Book of Why* by Pearl and Machenzie (2018)
- *Causal Inference* by Hernán and Robins (2020)
- *Dynamic Treatment Regimes* by Tsiatis et al. (2020)

But hopefully, being tailored toward pharmaceutical statistics, this book would be an introductory book for clinical statisticians to appreciate the importance of causal thinking in the planning, design, conduct, analysis, and interpretation of clinical studies. I select the topics to be presented in this book according to their importance and degree of interest in clinical studies, rather than according to their simplicity. This book is suitable as a textbook for a graduate or Master's level biostatistics course.

I would like to take this opportunity to thank many professors who taught me, including but not limited to, Professor Senlin Xu, Professor Lincheng Zhao, Professor Zhiliang Ying, and Professor Daniel Rabinowitz. In addition, I would like to thank my colleagues and friends for the many discussions that have inspired me. Most notably, I wish to thank Drs. Weili He and Martin Ho for founding the ASA Biopharm Section RWE Scientific Working Group. As a member, then co-lead, and then co-chair of this working group, I have benefited significantly from the group's many discussions with Drs. Jie Chen, Peng Ding, Susan Gruber, Hana Lee, Mark Levenson, Shu Yang, and Xiang Zhang, Professor Mark van der Laan, and other group members.

Finally, I am grateful to David Grubbs and Curtis Hill at Chapman & Hall/CRC Press for their great help and support.

1

Introduction

1.1 Central Questions

We start with a quote by John Tukey:

> "Far better an approximate answer to the right question, which is often vague, than the exact answer to the wrong question, which can always be made precise."—John Tukey

Then, what are the right questions we may ask as clinical statisticians working in the pharmaceutical industry? There are questions such as:

- What is the disease burden in patients with a disease?
- What is the treatment pattern of a patient population?
- What is the efficacy and safety of an investigational drug?

To find out what central questions for drug development and licensing are, we turn to guidance documents provided by the International Council for Harmonisation of Technical Requirements for Pharmaceuticals for Human Use (ICH), which can be downloaded from www.ich.org. Especially, we turn to guidance document ICH E9(R1) titled "Addendum on estimands and sensitivity analysis in clinical trials to the guideline on statistical principles for clinical trials." ICH E9(R1) posed the following questions:

> "Central questions for drug development and licensing are to establish the existence, and to estimate the magnitude, of treatment effects: how the outcome of treatment compares to what would have happened to the same subjects under alternative treatment (i.e. had they not received the treatment, or had they received a different treatment)."—ICH E9(R1)

Here is the anatomy of the above central questions:

- *To establish the existence, and to estimate the magnitude, of treatment effects*: Hypothesis testing, which is applied to establish the existence of the treatment effect, and confidence interval estimating, which is applied

DOI: 10.1201/9781003433378-1

to estimate the magnitude of the treatment effect, are the two most commonly used tools to conduct statistical inference. The target estimand in these statistical inferences is the treatment effect.

- *Treatment [compares to] alternative treatment*: Often the treatment is the investigational treatment or the treatment of interest, while the alternative treatment is the comparator, which may be a placebo, the standard of care, or a different treatment.
- *The same subjects*: These subjects form the population of subjects (e.g., patients or study participants) who would be treated by the treatment or the alternative treatment. How to imagine that the same subjects were to be treated by two different treatments (i.e., the treatment and the alternative treatment) relies on counterfactual thinking.
- *The outcome of treatment*: The outcome variable is some measurement of how a subject feels, functions, or survives. It is also known as response variable or endpoint.
- *What would have happened to the same subjects under alternative treatment*: This is the definition of potential outcome or counterfactual outcome, noting the subjunctive mood.
- *How the outcome of treatment compares to what would have happened to the same subjects under alternative treatment (i.e., had they not received the treatment, or had they received a different treatment)*: This is the definition of the treatment effect in terms of potential outcomes, noting the subjunctive mood.

Furthermore, ICH E9(R1) provided the definition of estimand:

> "An estimand is a precise description of the treatment effect reflecting the clinical question posed by a given clinical trial objective. It summarises at a population level what the outcomes would be in the same patients under different treatment conditions being compared."—ICH E9(R1)

From the above anatomy of central questions and definition of estimand, we see that ICH E9(R1) used the phrase "*what would have happened to the same subjects under alternative treatment*" to elucidate the term **counterfactual outcome** (Pearl 2009) and used the phrase "*what the outcomes would be in the same patients under different treatment conditions being compared*" to elucidate the term **potential outcomes** (Neyman 1923; Rubin 1975).

In the next section, we discuss potential and counterfactual outcomes from a philosophical point of view.

1.2 Potential Outcomes

From the anatomy of the central questions, we see that there are three basic constituent parts—still using the phrases in ICH E9(R1):

1. "The **population** of patients targeted by the clinical question."
2. "The **treatment** condition of interest and, as appropriate, the alternative treatment condition to which comparison will be made."
3. "The **variable** (or endpoint) to be obtained for each patient that is required to address the clinical question."

Let's supply some mathematical symbols for these three components. Let $i = 1, \ldots, N$ index the patients in the population, where N is the population size. In some studies, the subjects are healthy participants, but, for simplicity, hereafter we use "patients" instead of "patients/participants," and we may use "subjects" and "patients" interchangeably.

Let 1 indicate the treatment of interest and 0 indicate the control treatment (e.g., placebo, the standard of care, or alternative active treatment).

Let Y be the primary outcome variable, which could be continuous, binary, or time-to-event.

To fully understand the central questions, we rely on our ability of imagining, which is the top rung of the Ladder of Causation that consists of three rungs—seeing, doing, and imagining—described in *The Book of Why* (Pearl and Mackenzie 2018).

We imagine a world, referred to as the potential world, in which each subject possesses two potential outcomes before any study is conducted and any outcome is measured. This imagination is expressed by the subjunctive mood in ICH E9(R1).

The imagination of the potential world is not without constraint. For example, one constraint is **the stable unit treatment value assumption** (SUTVA) (Rubin 1980): "The potential outcomes for any unit do not vary with the treatments assigned to other units, and, for each unit, there are no different forms or versions of each treatment level, which lead to different potential outcomes."

Let Y_i^a be the potential outcome that would happen to patient i if the patient were treated by treatment a, where $a = 1$ or 0. That is, in the potential world satisfying SUTVA, the population consists of N patients, with each patient carrying two potential outcomes, Y_i^1 and Y_i^0. Under SUTVA, $(Y_i^1, Y_i^0), i = 1, \ldots, N$, are independent.

In the real world, patient i is treated by treatment A_i, which is either 1 or 0. Let Y_i be the corresponding outcome after being treated by A_i. Thus, $Y_i, i = 1, \ldots, N$, are observable outcomes in the real world.

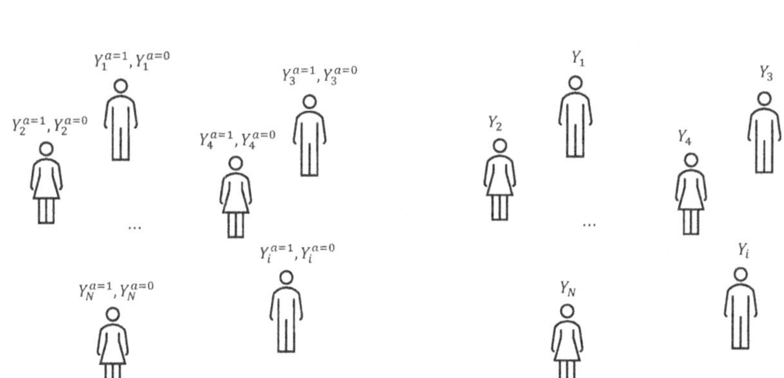

FIGURE 1.1
The two worlds—the potential world and the real world

From a philosophical point of view, the potential world is like Platonic world of ideas, which can be understood by Plato's allegory of the cave[1].

Using Plato's allegory of the cave, we can think of the observed data $\{Y_i, i = 1, \ldots, N\}$ that exist in the real world as a mere shadow of the full data $\{(Y_i^1, Y_i^0), i = 1, \ldots, N\}$ that exist in the potential world. See Figure 1.1 for an illustration of the two worlds.

An estimand is a quantity that is to be estimated in a study, which will be the target of the study throughout study planning, design, conduct, analysis, and interpretation stages. An estimand defined in terms of potential outcomes, like the one defined in ICH E9(R1), is referred to as *causal estimand*, which has causal meaning. Causal inference is the process of estimating the causal estimand based on the observed data. In defining the causal estimand, we are taking the Platonic viewpoint, while in doing causal inference, we are taking the Aristotelian viewpoint—using the observable data of the real world to infer a quantity of the potential world.

However, whether we are able to infer a quantity of the potential world using the data from the real world depends on an implicit assumption, the **consistency assumption** (Hernán and Robins 2020), which assumes that the real world is one realization of the potential world. Without assuming the consistency assumption (i.e., if the consistency assumption is violated),

1. For Plato's allegory of the case, see an excerpt from *Sophie's World* (Gaarder 1994): "Imagine some people living in an underground cave. They sit with their backs to the mouth of the cave with their hands and feet bound in such a way that they can only look at the back wall of the cave. Behind them is a high wall, and behind that wall pass human-like creatures, holding up various figures above the top of the wall. Because there is a fire behind these figures, they cast flickering shadows on the back wall of the cave. So the only thing the cave dwellers can see is this shadow play. They have been sitting in this position since they were born, so they think these shadows are all there are."

the two worlds would be disconnected, causing that we couldn't infer any quantity of the potential world based on the data in the real world.

We can express the consistency assumption explicitly. Recall that in the real world subject i receives treatment A_i and has outcome Y_i. The consistency assumption states that the observed outcome is consistent with one of the two potential outcomes, the one associated with the value of A_i. That is, if $A_i = 1$, then $Y_i = Y_i^1$; if $A_i = 0$, then $Y_i = Y_i^0$. In short, the consistency assumption assumes that $Y_i = Y_i^{A_i}$, $i = 1, \ldots, N$.

Furthermore, let $B_i = 1 - A_i$. That is, if $A_i = 1$, then $B_i = 0$; if $A_i = 0$, then $B_i = 1$. Hence, B_i is the alternative treatment for subject i given that subject i is treated by treatment A_i. In the potential world, Y_i^a is the potential outcome associated with $A_i = a$ and Y_i^b is the potential outcome associated with $B_i = b$. In the real world, Y^{A_i} is the factual outcome given that subject i is treated by treatment A_i actually, while Y^{B_i} is the counterfactual outcome given that subject i is not treated by the alternative treatment B_i actually.

1.3 Estimand

1.3.1 The PROTECT checklist

As discussed in the previous section, the central questions are posed in a tripartite form—population, treatment, and outcome—via the counterfactual thinking. However, another key component, which is **time**, has not yet been discussed. Time plays an important role in specifying both treatment and outcome (e.g., when the baseline is defined, when the treatment is initiated, and when the outcome is measured).

Figure 1.2 illustrates the time interval between the baseline denoted as $t = 0$, the time when the treatment is initiated, and the endpoint denoted as $t = T$, the time when the outcome is measured—this is why outcome variables are also known as endpoints in clinical studies. Variables that are measured at the baseline are often referred to as baseline characteristics or covariates; if the baseline is defined as the time of treatment initiation, such variables are also known as pre-treatment variables.

Moreover, we should identify the type of the outcome variable. The three most common types in clinical studies are: continuous outcome, binary outcome, and time-to-event outcome. If the outcome variable is continuous, for each subject, it takes on a quantitative value at time T. If the outcome variable is binary, for each subject, it takes on one of the two values (say, 0 for success and 1 for failure) at time T. If the outcome variable is time-to-event, it is subject to being censored. Since censoring is a challenging problem, most of time we will concentrate the discussion on continuous and binary outcomes, while we will discuss time-to-event outcomes whenever it is ready.

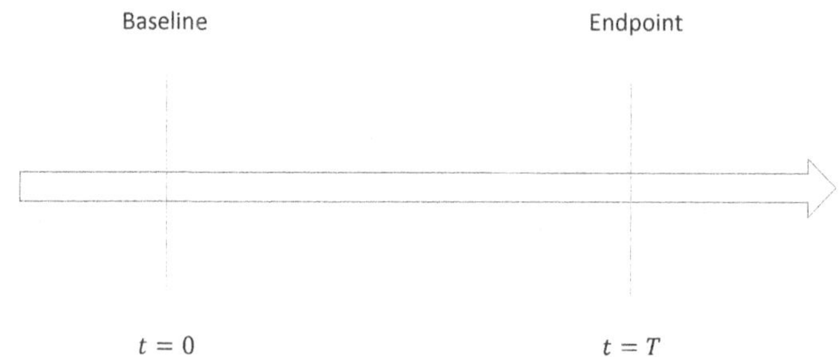

FIGURE 1.2
Two time-points in a typical clinical study—baseline and endpoint

After discussing the above five components (population, outcome, treatment, counterfactual thinking, and time), we have a checklist for specifying a sound research question and, consequently, specifying a study objective and defining a target estimand. The checklist, referred to as the PROTECT checklist (Fang, Wang, and He 2020), is summarized in Table 1.1.

1.3.2 Estimand for a given population

After gathering those items according to the PROTECT checklist, for a given population $\mathcal{P} = \{1, \ldots, N\}$, we can define an estimand that reflects the research question. Here are three examples that are corresponding to three types of outcome variables—continuous, binary, and time-to-event, respectively.

If the outcome variable is continuous, for patient i, let Y_i^1 and Y_i^0 be the two quantitative measurements that would be measured in the potential world at time T after baseline, were the patient to be treated by treatment 1 and 0, respectively. Then $Y_i^1 - Y_i^0$ measures the *individual treatment effect* for patient i (Imbens and Rubin 2015). To interpret this, assume that Y is some health measurement that is the bigger the better. Thus, for patient i, $Y_i^1 - Y_i^0 > 0$ means treatment 1 is more beneficial than treatment 0 and vice versus, and $Y_i^1 - Y_i^0 = 0$ means there is no treatment effect. Hence, the following **average**

TABLE 1.1
The PROTECT checklist.

Symbol	Checklist	Explanation
P	Population	Population defined via I/E criteria*
R/O	Response/Outcome	Primary dependent variable
T/E	Treatment/Exposure	Primary independent variable
C	Counterfactual thinking	Potential or counterfactual outcomes
T	Time	When the variables are measured?

* I/E criteria are inclusion/exclusion criteria

treatment effect (ATE) is an estimand of interest:

$$\theta(\mathcal{P}) = \frac{1}{N}\sum_{i=1}^{N}(Y_i^1 - Y_i^0). \tag{1.1}$$

Here, the population mean is the population-level summary that is used to summarize the treatment effect. Although other summaries such as median are also applicable, we consider mean for statistical convenience.

If the outcome is binary, although any coding system is applicable, we use 1/0 coding for statistical convenience (e.g., 1 for failure or 0 for success). Let Y_i^1 and Y_i^0 be the two potential outcomes for patient i. If 1/0 means failure/success, then for patient i, $(Y_i^1, Y_i^0) = (0, 1)$ means treatment 1 is beneficial, $(Y_i^1, Y_i^0) = (1, 0)$ means treatment 1 is harmful, and $(Y_i^1, Y_i^0) = (1, 1)$ or $(0, 0)$ means there is no treatment effect. Thus, the following ATE is an estimand of interest:

$$\theta(\mathcal{P}) = \frac{1}{N}\sum_{i=1}^{N}\left\{I(Y_i^1 = 1) - I(Y_i^0 = 1)\right\}, \tag{1.2}$$

where $I(\cdot)$ is the indicator function. In this example, the population proportion is the population-level summary, noting that, for 1/0–coded binary variable, mean is equivalent to proportion.

If the outcome is time-to-event (e.g., time to death starting from baseline), for patient i, let Y_i^1 and Y_i^0 be the two potential outcomes, were the patient to be treated by treatment 1 and 0, respectively. Thus, if we consider mean survival time as the population-level summary, we can define the same ATE estimand as in (1.1), noting that time-to-event outcome is one special case of continuous outcome if there is no censoring. Although estimand like (1.1) for the time-to-event outcome is well-defined in the potential world, this estimand is seldom considered in clinical studies, because often it is too time-consuming, if not impossible, to keep the study running until all the patients observe the event of interest. Therefore, we may consider the following two alternative population-level summaries in the estimand definition.

One alternative summary is restricted-mean-survival-time (RMST) limited by T (Royston and Parmar 2011), leading to the following estimand of interest:

$$\theta'(\mathcal{P}) = \frac{1}{N}\sum_{i=1}^{N}\left[\min\{Y_i^1, T\} - \min\{Y_i^0, T\}\right]. \tag{1.3}$$

Another alternative summary is survival rate at T (Kalbfleisch and Prentice 2011), leading to the following estimand of interest:

$$\theta''(\mathcal{P}) = \frac{1}{N}\sum_{i=1}^{N}\left\{I(Y_i^1 > T) - I(Y_i^0 > T)\right\}. \tag{1.4}$$

1.3.3 Estimand for a given super-population

To simplify the construction of the aforementioned estimands, the concept of super-population has been introduced (Deming and Stephan 1941; Hernán and Robins 2020; Imbens and Rubin 2015).

We start with the empirical distribution function of N pairs of potential outcomes in the population $\mathcal{P}$. Let P_N be the empirical distribution, with probability mass $1/N$ on each pair of potential outcomes, (Y_i^1, Y_i^0), $i = 1, \ldots, N$. Then all the aforementioned estimands can rewritten in terms of P_N. For example, estimand (1.1) can be rewritten as

$$\theta(\mathcal{P}) = \mathbb{E}_{(Y^1,Y^0)\sim P_N}\left(Y^1 - Y^0\right), \tag{1.5}$$

where, by convention, variables without subscript i are the ones of a subject who is randomly selected from a given population—here the given population is the population equipped with the distribution P_N.

Now we are ready to imagine a super-population. For this aim, we conceptually increase the size of the population $\mathcal{P}$, increasing it from N to infinitely large. Then the resulting population is called the super-population. Let $P_\infty = \lim_{N\to\infty} P_N$. Thus, P_∞ would be the probability distribution function of the pair of potential outcomes, (Y^1, Y^0), of a patient who were randomly selected from the super-population.

Finally, we are able to define the estimand of interest associated with the super-population equipped with the distribution P_∞. For example, for continuous outcome, if the estimand of interest for the population $\mathcal{P}$ equipped with the distribution P_N is (1.5), then the estimand of interest for the super-population equipped with the distribution P_∞ is

$$\theta^* = \mathbb{E}_{(Y^1,Y^0)\sim P_\infty}\left(Y^1 - Y^0\right). \tag{1.6}$$

1.3.4 Internal validity and external validity

In a given study, a **target population** is defined via a set of inclusion/exclusion criteria. In other words, the target population consists of all the patients who live on the earth and are eligible to be enrolled in the clinical study, meeting the inclusion/exclusion criteria.

After one study is completed, there are N subjects who are truly enrolled in the study. However, these subjects are usually enrolled by convenience from a number of enrollment sites, except that in some survey studies the subjects are sampled randomly from the target population. Now the question is: Does this group of N enrolled subjects represent the target population if the subjects are sampled by convenience? The answer is no. That is why we need to distinguish between internal validity and external validity.

On the one hand, the group of N subjects with full data $\{(Y_i^1, Y_i^0), i = 1, \ldots, N\}$ can be considered as a **population**. This population of size N is called the intent-to-treat (ITT) population—because we intend to treat them by either treatment 1 or treatment 0. Earlier we denote this population as $\mathcal{P}$. According to the mental gymnastics of constructing a super-population, we can conceptually construct a **super-population** such that $(Y_i^1, Y_i^0), i = 1, \ldots, N$, are N independent and identically distributed (i.i.d.) subjects drawn from the super-population. If the estimand of interest is associated with the super-population, say θ^* in (1.6), then we attempt to use the observed data of N subjects in $\mathcal{P}$ to estimate θ^*. In such attempt, we encounter the issue of internal validity.

On the other hand, the same group of N subjects can also be considered as a **sample** that is sampled by convenience from the target population, denoted as $\mathcal{T}$, which is defined via the set of inclusion/exclusion criteria. In a similar way that we define $\theta(\mathcal{P})$, we can define $\theta(\mathcal{T})$. If $\theta(\mathcal{T})$ is the estimand of interest, then we attempt to use the observed data of N subjects in $\mathcal{P}$ to estimate it. In such an attempt, we encounter the issue of external validity. External validity is to examine the extent to which we could generalize the results we obtain from the sample of N subjects to the target population $\mathcal{T}$.

To summarize, for the group of N subjects, there is a **sample–population duality**, similar to the wave–particle duality of light. On the one hand, to examine the internal validity, the group of N subjects is a population with full data $\{(Y_i^1, Y_i^0), i = 1, \ldots, N\}$ that are conceptually drawn from the super-population. On the other hand, to examine the external validity, the group of N subjects is a sample that is sampled by convenience from the target population. This sample–population duality is illustrated in Figure 1.3.

The remaining of this book will focus on the internal validity; that is, the estimand of interest is defined for a given super-population from which a group of N i.i.d. subjects are drawn. If we want to further examine the external validity, we should collect external data to examine how well $\mathcal{P}$ represents $\mathcal{T}$.

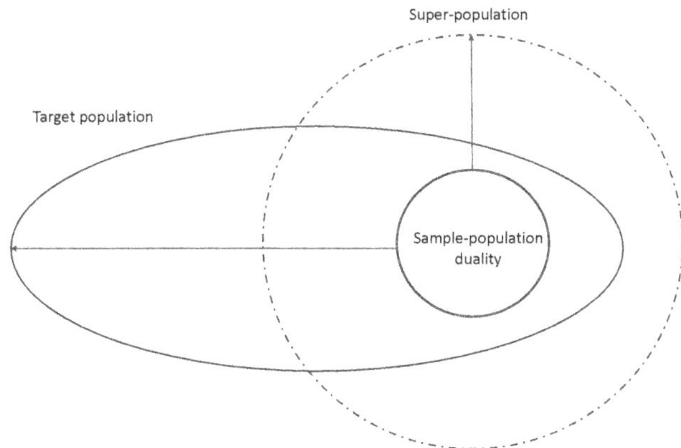

FIGURE 1.3
The sample–population duality

1.4 Probability and Statistics

1.4.1 Probability

In this subsection, we review some basic probability results, which will be used in the following chapters and which can be found in any introductory textbook of probability theory; e.g., Chung (2001).

Probability distribution

For a discrete random variable X, let $p_X(x) = \mathbb{P}(X = x)$ be its probability mass function and $F_X(x) = \mathbb{P}(X \leq x)$ be its cumulative distribution function,

$$F_X(x) = \sum_{x' \leq x} p_X(x').$$

For a continuous random variable X, let $p_X(x)$ be its probability density function and $F_X(x) = \mathbb{P}(X \leq x)$ be its cumulative distribution function,

$$F_X(x) = \int_{-\infty}^{x} p_X(x')dx'.$$

To abuse the notation, we use the same notation, $p_X(x)$, for both probability mass function and probability density function. For simplicity, most of time we state the results in terms of probability mass functions for discrete random variables, but all the statements for discrete random variables to be reviewed here are also true for continuous variables, except that summation is replaced by integration.

Joint, marginal, and conditional distributions

Let X and Y be two discrete random variables, with joint distribution $p_{X,Y}(x,y) = \mathbb{P}(X = x, Y = y)$. Marginal distributions of X and Y are $p_X(x) = \sum_y p_{X,Y}(x,y)$ and $p_Y(y) = \sum_x p_{X,Y}(x,y)$, respectively. Conditional distribution of Y given $X = x$ if $p_X(x) \neq 0$ is

$$p_{Y|x}(y|x) = \frac{p_{X,Y}(x,y)}{p_X(x)}.$$

Thus, $p_{X,Y}(x,y) = p_X(x)p_{Y|x}(y|x)$, referred to as the chain rule. The chain rule holds if the order of X and Y is changed; i.e., $p_{X,Y}(x,y) = p_Y(y)p_{X|y}(x|y)$. By induction, for any $X_1, \dots, X_k$, we can show the following chain rule,

$$p_{X_1,\dots,X_k}(x_1,\dots,x_k) = p_{X_1}(x_1)p_{X_2|x_1}(x_2|x_1)\cdots p_{X_k|x_1,\dots,x_{k-1}}(x_k|x_1,\dots,x_{k-1}).$$

Independence and conditional independence

If $p_{X|y}(x|y) = p_X(x)$ for any x and y, then we say X and Y are independent, which is denoted as $X \perp\!\!\!\perp Y$. By the definition of conditional probability, we have an equivalent definition of independence: if $p_{X,Y}(x, y) = p_X(x)p_Y(y)$ for any x and y, then $X \perp\!\!\!\perp Y$.

We can also define conditional independence where there is another variable, say Z. If $p_{X,Y|z}(x, y|z) = p_{X|z}(x|z)p_{Y|z}(y|z)$ for any x, y, and z, then X and Y are conditionally independent given Z, denoted as $X \perp\!\!\!\perp Y|Z$.

Expectation, variance, and covariance

Expectation of X is defined as

$$\mathbb{E}(X) = \sum_x x p_X(x) = \int x dF_X(x).$$

Variance of X is defined as

$$\mathbb{V}(X) = \mathbb{E}[X - \mathbb{E}(X)]^2 = \mathbb{E}(X^2) - [\mathbb{E}(X)]^2.$$

Covariance X and Y is defined as

$$\text{Cov}(X, Y) = \mathbb{E}[X - \mathbb{E}(X)][Y - \mathbb{E}(Y)] = \mathbb{E}(XY) - \mathbb{E}(X)\mathbb{E}(Y).$$

Law of iterated expectations

The law of iterated expectations states that

$$\mathbb{E}[\mathbb{E}(X|Y)] = \mathbb{E}(X).$$

Proof of the law of iterated expectation:

$$\begin{aligned}
\mathbb{E}[\mathbb{E}(X|Y)] &= \sum_y \mathbb{E}(X|Y = y)p_Y(y) \\
&= \sum_y \sum_x x p_{X|y}(x|y)p_Y(y) \\
&= \sum_y \sum_x x p_{X,Y}(x, y) \\
&= \sum_x x \sum_y p_{X,Y}(x, y) \\
&= \sum_x x p_X(x) = \mathbb{E}(X). \qquad \square
\end{aligned}$$

By the law of iterated expectation, we can show that

$$\mathbb{V}(X) = \mathbb{E}[\mathbb{V}(X|Y)] + \mathbb{V}[\mathbb{E}(X|Y)].$$

Convergence

A sequence of random variables, $X_1, X_2, \ldots$, converges in probability to random variable X if for any $\varepsilon > 0$,

$$\lim_{n\to\infty} \mathbb{P}(|X_n - X| \geq \varepsilon) = 0.$$

We denote the convergence in probability as

$$X_n \xrightarrow{p} X \quad \text{or} \quad X_n = X + o_p(1).$$

A sequence of random variables, $X_1, X_2, \ldots$, converges in distribution to random variable X if for any x at which $F_X(x)$ is continuous,

$$\lim_{n\to\infty} F_{X_n}(x) = F_X(x).$$

We denote the convergence in distribution as

$$X_n \xrightarrow{d} X.$$

A theorem that involves both convergence in probability and convergence in distribution is known as Slutsky's lemma: If $X_n \xrightarrow{d} X$ and $Y_n \xrightarrow{p} c$, then

$$\begin{aligned} &(i) \quad X_n + Y_n \xrightarrow{d} X + c; \\ &(ii) \quad Y_n X_n \xrightarrow{d} cX; \\ &(iii) \quad Y_n^{-1} X_n \xrightarrow{d} c^{-1} X, \text{ if } c \neq 0. \end{aligned}$$

Law of large numbers

Let $X_1, X_2, \ldots$ be a sequence of i.i.d. random variables with $\mathbb{E}(X_1) = \mu < \infty$. The law of large numbers states that

$$\frac{1}{n}\sum_{i=1}^{n} X_i \xrightarrow{p} \mu.$$

Central limit theorem

Let $X_1, X_2, \ldots$ be a sequence of i.i.d. random variables with $\mathbb{E}(X_1) = \mu < \infty$ and $\mathbb{V}(X_1) = \sigma^2 < \infty$. The central limit theorem states that

$$\sqrt{n}\left(\frac{1}{n}\sum_{i=1}^{n} X_i - \mu\right) \xrightarrow{d} \mathcal{N}(0, \sigma^2),$$

where $\mathcal{N}(0, \sigma^2)$ is a normal distribution with mean zero and variance σ^2.

1.4.2 Directed acyclic graphs

In this subsection, we review some brief results of directed acyclic graphs (DAGs), which will be used in the following chapters. Please refer to Pearl, Glymour, and Jewell (2016), Greenland, Pearl, and Robins (1999), or Dawid (2010) for a brief review and refer to Pearl (2009) or Hernán and Robins (2020) for a comprehensive review.

Figure 1.4 displays three simple DAGs—chain, fork, and collider in (a)-(c) respectively—showing three possible relationships among three variables. These three basic DAGs may appear as sub-DAGs in any DAG; for example, DAG displayed in Figure 1.4(d) has a fork $A \leftarrow X \rightarrow Y$, a chain $X \rightarrow Y \rightarrow E$, and a collider E in $A \rightarrow E \leftarrow Y$.

Chain

The chain in Figure 1.4(a) displays the following relationships:

- A and M are likely dependent;
- M and Y are likely dependent;
- A and Y are likely dependent;
- Conditional on M, A and Y are independent.

Proof of $A \perp\!\!\!\perp Y|M$ in the chain: Let $p_{A,M,Y}(a, m, y)$ be the joint distribution of (A, M, Y). By the chain rule, $p_{A,M,Y}(a, m, y) = p_A(a)p_{M|a}(m|a)p_{Y|m}(y|m)$.

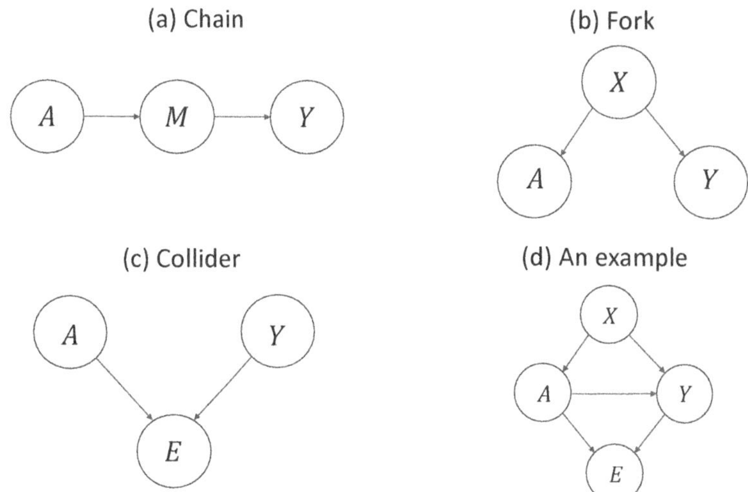

FIGURE 1.4
Chain, fork, collider, and an example

Thus,

$$\begin{aligned} p_{A,Y|m}(a,y|m) &= p_{A,M,Y}(a,m,y)/p_M(m) \\ &= p_A(a)p_{M|a}(m|a)p_{Y|m}(y|m)/p_M(m) \\ &= 1/p_M(m) \cdot p_A(a)p_{M|a}(m|a) \cdot p_{Y|m}(y|m). \end{aligned}$$

Since $p_{A,Y|m}(a,y|m)$ is the product of one function depending on a only and the other function on y only, we show that $A \perp\!\!\!\perp Y|M$. □

Fork

The fork in Figure 1.4(b) displays the following relationships:

- A and X are likely dependent;
- Y and X are likely dependent;
- A and Y are likely dependent;
- Conditional on X, A and Y are independent.

Proof of $A \perp\!\!\!\perp Y|X$ in the fork: Let $p_{X,A,Y}(x,a,y)$ be the joint distribution of (X,A,Y). By the chain rule, $p_{X,A,Y}(x,a,y) = p_X(x)p_{A|x}(a|x)p_{Y|x}(y|x)$. Thus,

$$\begin{aligned} p_{A,Y|x}(a,y|x) &= p_{X,A,Y}(x,a,y)/p_X(x) \\ &= p_X(x)p_{A|x}(a|x)p_{Y|x}(y|x)/p_X(x) \\ &= p_{A|x}(a|x) \cdot p_{Y|x}(y|x). \end{aligned}$$

Since $p_{A,Y|x}(a,y|x)$ is the product of one function depending on a only and the other function on y only, we show that $A \perp\!\!\!\perp Y|X$. □

Collider

The collider in Figure 1.4(c) display the following relationships:

- A and E are likely dependent;
- Y and E are likely dependent;
- A and Y are independent;
- Conditional on E, A and Y are likely dependent.

Proof of $A \not\perp\!\!\!\perp Y|E$ in the collider: Let $p_{A,Y,E}(a,y,e)$ be the joint distribution of (A,Y,E). By the chain rule, $p_{A,Y,E}(a,y,e) = p_A(a)p_Y(y)p_{E|a,y}(e|a,y)$. Thus,

$$\begin{aligned} p_{A,Y|e}(a,y|e) &= p_{A,Y,E}(a,y,e)/p_E(e) \\ &= p_A(a)p_Y(y)p_{E|a,y}(e|a,y)/p_E(e). \end{aligned}$$

Since $p_{E|a,y}(e|a,y)$ depends on both a and y, which cannot be written as a product of one function depending on a only and the other function on y only, we show that $A \not\perp\!\!\!\perp Y|E$. □

1.4.3 Statistics

In this subsection, we review some important results in asymptotic statistics, which will be used in the following chapters and which can be found in any textbook of asymptotic statistics; e.g., van der Vaart (2000).

Unbiasedness

Let $O_1, O_2, \ldots$ be a sequence of i.i.d. observations with common distribution P. Let $\theta = \theta(P)$ be the estimand of interest, which is a function of distribution P. Consider an estimator $\widehat{\theta}_n = \widehat{\theta}(O_1, \ldots, O_n)$, which is a function of data $\{O_1, \ldots, O_n\}$. We say that $\widehat{\theta}_n$ is an unbiased estimator of θ if

$$\mathbb{E}_P(\widehat{\theta}_n) = \theta,$$

where the subscript "P" means the expectation is taken over distribution P.

In particular, if the true distribution is denoted as P_0, then $\theta_0 = \theta(P_0)$ is the true value of θ and

$$\mathbb{E}(\widehat{\theta}_n) = \mathbb{E}_{P_0}(\widehat{\theta}_n) = \theta_0,$$

where we suppress the subscript P_0 if the expectation is taken over P_0.

Consistency

We say that $\widehat{\theta}_n$ is a consistent estimator of θ if

$$\widehat{\theta}_n \xrightarrow{p} \theta,$$

as $n \to \infty$ under any P. In particular, $\widehat{\theta}_n \xrightarrow{p} \theta_0$ under P_0.

The consistency defined here is an asymptotic property, which should not be confused with the consistency assumption—an identifiability assumption.

Asymptotic normality

We say that $\widehat{\theta}_n$ is asymptotically normal if

$$\sqrt{n}\left(\widehat{\theta}_n - \theta(P)\right) \xrightarrow{d} \mathcal{N}\left(0, \sigma^2(P)\right),$$

with asymptotic variance $\sigma^2(P)$, as $n \to \infty$ under any P. In particular, $\sqrt{n}(\widehat{\theta}_n - \theta_0) \xrightarrow{d} \mathcal{N}(0, \sigma_0^2)$ under P_0, where $\sigma_0^2 = \sigma^2(P_0)$.

Z-statistic

If we can show the consistency and asymptotic normality of $\widehat{\theta}_n$ and construct a consistent estimator—denoted as $\widehat{\sigma}_n^2$— for its asymptotic variance (i.e., $\widehat{\sigma}_n^2 \xrightarrow{p} \sigma_0^2$), then we can construct the Z-statistic,

$$Z_n = \frac{\sqrt{n}(\widehat{\theta}_n - \theta_0)}{\widehat{\sigma}_n},$$

leading to $Z_n \xrightarrow{d} \mathcal{N}(0,1)$ by Slutsky's lemma.

Confidence interval estimator based on Z-statistic

We can construct a 95% confidence interval estimator for θ_0:

$$(\widehat{\theta}_n - 1.96 \times \widehat{\sigma}_n/\sqrt{n},\ \widehat{\theta}_n + 1.96 \times \widehat{\sigma}_n/\sqrt{n}),$$

which can be written as $\widehat{\theta}_n \pm 1.96 \times \widehat{\text{SE}}$, where $\widehat{\text{SE}} = \widehat{\sigma}_n/\sqrt{n}$ is an estimator of the standard error (SE) of $\widehat{\theta}_n$.

Hypothesis testing based on Z-test

We can conduct hypothesis testing for the following pair of hypothesis:

$$H_0 : \theta = \theta_{\text{null}} \quad \text{vs.} \quad H_a : \theta \neq \theta_{\text{null}},$$

where θ_{null} is a given value of θ that is specified in the null hypothesis. For example, if $\theta_{\text{null}} = 0$, then the null hypothesis states that there is no treatment effect. The two-sided Z-test statistic is defined as

$$z^{\text{obs}} = \frac{\sqrt{n}(\widehat{\theta}_n - \theta_{\text{null}})}{\widehat{\sigma}_n},$$

and therefore the two-sided p-value is equal to $2\mathbb{P}\{\mathcal{N}(0,1) \geq |z^{\text{obs}}|\}$.

Asymptotic linearity

A special consistent and asymptotic normal estimator is asymptotically linear estimator. We say that $\widehat{\theta}_n$ is asymptotically linear if

$$\sqrt{n}(\widehat{\theta}_n - \theta_0) = \frac{1}{\sqrt{n}} \sum_{i=1}^{n} \phi(O_i) + o_p(1), \tag{1.7}$$

where $\phi(O)$ is a function of observation O with $\mathbb{E}(\phi(O)) = 0$ and $\mathbb{V}(\phi(O)) < \infty$. Here $\phi(O)$ is called the *influence function* associated with estimator $\widehat{\theta}_n$.

We can easily see that the asymptotic linearity implies the consistency (thanks to the law of large numbers) and the asymptotic normality (thanks to the central limit theorem).

1.5 Exercises

Ex 1.1

Download guidance documents ICH E9 "Statistical principles for clinical trials" and ICH E9(R1) "Addendum on estimands and sensitivity analysis in clinical trials" from the ICH official website, www.ich.org. There are four groups of guidance documents: quality guidelines, safety guidelines, efficacy guidelines, and multidisciplinary guidelines. Read the introduction section of ICH E9 and the first three sections of ICH E9(R1).

Ex 1.2

In R, generate a data frame that has 4 columns (subject ID `SID`, treatment variable `A` being 0 or 1, potential outcome `Y0`, and potential outcome `Y1`) and 100 rows, using the following R codes:

```
set.seed(1)
SID <- 1:100
A <- rbinom(n = 100, size = 1, prob = 0.5)
Y0 <- rnorm(n = 100, mean = 0, sd = 1)
Y1 <- rnorm(n = 100, mean = 1, sd = 1.5)
dataset1 <- data.frame(SID = SID, A = A, Y0 = Y0, Y1 = Y1)
```

Based on the above population of size $N = 100$ that exists in the potential world, what is the value of the estimand defined in (1.1)?

Ex 1.3

In R, generate a data frame that has 4 columns (subject ID `SID`, treatment variable `A` being 0 or 1, potential outcome `Y0`, and potential outcome `Y1`) and 100 rows, using the following R codes:

```
set.seed(2)
SID <- 1:100
A <- rbinom(n = 100, size = 1, prob = 0.5)
Y0 <- rbinom(n = 100, size = 1, prob = 0.3)
Y1 <- rbinom(n = 100, size = 1, prob = 0.7)
dataset2 <- data.frame(SID = SID, A = A, Y0 = Y0, Y1 = Y1)
```

Based on the above population of size $N = 100$ that exists in the potential world, what is the value of the estimand defined in (1.2)?

Ex 1.4

Let $X_1, X_2, \ldots$ be a sequence of i.i.d. continuous random variables with common distribution P. Let $\theta_0 = \mathbb{E}(X_1)$. Assume that $\mathbb{V}(X_1) < \infty$. Consider the following estimator—the sample mean:

$$\widehat{\theta}_n = \frac{1}{n} \sum_{i=1}^{n} X_i.$$

Verify that $\widehat{\theta}_n$ is an unbiased, consistent, asymptotically normal, and asymptotically linear estimator of θ_0. What is the influence function associated with $\widehat{\theta}_n$ for estimating θ_0?

Ex1.5

Let $X_1, X_2, \dots$ be a sequence of i.i.d. 1/0–coded binary random variables with $\theta_0 = \mathbb{P}(X_1 = 1)$. Consider the following estimator—the sample proportion:

$$\widehat{\theta}_n = \frac{1}{n} \sum_{i=1}^{n} I(X_i = 1).$$

Verify that $\widehat{\theta}_n$ is an unbiased, consistent, asymptotically normal, and asymptotically linear estimator of θ_0. What is the influence function associated with $\widehat{\theta}_n$ for estimating θ_0?

2

Randomized Controlled Clinical Trials

2.1 Randomization and Blinding

In *Fisher, Bradford Hill, and Randomization* (Armitage 2003), the contribution of two great statisticians to randomization was summarized:

> "In the 1920s, RA Fisher presented randomization as an essential ingredient of his approach to the design and analysis of experiments, validating significance tests. [...] Twenty years later, A Bradford Hill promulgated the random assignment of treatments in clinical trials as the only means of avoiding systematic bias between the characteristics of patients assigned to different treatments."—Armitage (2003)

To understand why "the random assignment of treatments" is "the only means of avoiding systematic bias," we supply some mathematical symbols. Consider a population of N subjects indexed by i. Let Z_i be the treatment assignment for subject i to either treatment 1 or treatment 0, and let Y_i be the outcome variable measured at time T after the treatment initiation, $i = 1, \ldots, N$.

The left panel of Figure 2.1 shows the directed acyclic graph (DAG) depicting relationship between Z and Y, following the convention that variable names without subscript i are for a subject randomly selected from the population. If the treatment assignment Z were to be determined by the subjects themselves or by their doctors, there would be an underlying variable or an underlying vector of variables (denoted as U) related to both the treatment assignment and the outcome. Then the relation between Z and Y would be the combination of the treatment effect of Z on Y and the confounding effect due to common cause U. Therefore, even if the causal relation between Z and Y were null, there would be an association between Z and Y because of the confounding effect due to U.

However, if the treatments are randomly assigned to subjects, Z does not depend on any underlying variables that are associated with Y. That is, by randomization, we cut off the arrow from U to Z in the left-panel of

DOI: 10.1201/9781003433378-2

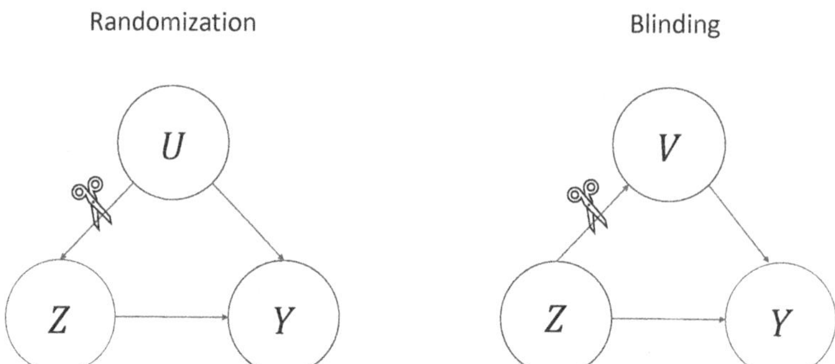

FIGURE 2.1
The use of randomization and blinding

Figure 2.1. By this means, we avoid the confounding bias. We refer to clinical trials with randomization as randomized controlled clinical trials (RCTs).

There are two types of RCTs: open-label RCTs and double-blind RCTs. In an open-label RCT, both the investigators and the patients know the treatment assignments. Consequently, by knowing the treatment assignments, the investigators may treat the patients differently consciously or unconsciously, and the knowledge of treatment assignments may affect the patients' behaviors and responses in the study.

The right-panel of Figure 2.1 displays the DAG illustrating that the total effect of Z on Y in an open-label RCT may consist of two effects: (1) the direct effect of Z on Y depicted by path $Z \to Y$ and (2) the indirect effect of Z on Y through V depicted by path $Z \to V \to Y$.

By designing a double-blind RCT, we cut off the arrow from Z to V in the right-panel of Figure 2.1. By means of blinding, the total effect Z on Y is equal to the direct effect of Z on Y. In double-blind RCTs, the comparator that is indistinguishable from the investigative treatment is called a placebo.

The guidance document ICH E9 "Statistical principles for clinical trials" emphasized the importance of randomization and blinding as follows:

> "The most important design techniques for avoiding bias in clinical trials are blinding and randomisation, and these should be normal features of most controlled clinical trials intended to be included in a marketing application."—ICH E9

To summarize, in a RCT, by the means of randomization, we avoid the confounding bias due to path $Z \leftarrow U \to Y$ and make it possible to estimate the effect of Z on Y unbiasedly. In a double-blind RCT, by the means of blinding, we remove the indirect effect due to path $Z \to V \to Y$ and the total effect of Z on Y is the same as the direct effect of Z on Y.

2.2 Estimand

2.2.1 Causal estimand

Continue the discussion of potential outcomes in Chapter 1. In the potential world, before the study is conducted, subject i would have potential outcomes:

$Y_i^{z=1}$ if subject i were assigned to treatment 1,
$Y_i^{z=0}$ if subject i were assigned to treatment 0,
$Y_i^{a=1}$ if subject i were treated by treatment 1,
$Y_i^{a=0}$ if subject i were treated by treatment 0.

Note that there could be any number of potential outcomes if we keep imagining feasible interventions to be put on the superscript of Y_i. In Chapter 1 we imagine two potential outcomes, while here we imagine four potential outcomes. We can imagine more potential outcomes provided that there are more counterfactual interventions or actions of interest.

Therefore, in the potential world, we imagine that the population $\mathcal{P}$ would consist of $(Y_i^{z=1}, Y_i^{z=0}, Y_i^{a=1}, Y_i^{a=0}, \dots), i = 1, \dots, N$, which are assumed to be satisfying the stable unite treatment value assumption (SUTVA), before the study is conducted. In clinical studies, this population is called the the intent-to-treat (ITT) population. Furthermore, recall that we can conceptually construct a super-population such that these N sets of potential outcomes are i.i.d. drawn from the super-population.

In this chapter, we only use the pairs of potential outcomes $\{(Y_i^{z=1}, Y_i^{z=0})\}$ to define estimands; while in the next chapter, we will consider the quartets of potential outcomes $\{(Y_i^{z=1}, Y_i^{z=0}, Y_i^{a=1}, Y_i^{a=0})\}$, when we discuss the realistic compliance problem.

The outcome variable may be continuous or binary. For continuous outcome, we may consider mean as the population-level summary and define the following estimand associated with the population $\mathcal{P}$,

$$\theta(\mathcal{P}) = \frac{1}{N} \sum_{i=1}^{N} \left(Y_i^{z=1} - Y_i^{z=0} \right), \tag{2.1}$$

which is called the ITT estimand. Thus, the ITT estimand associated with the super-population is defined as

$$\theta^* = \mathbb{E} \left(Y^{z=1} - Y^{z=0} \right). \tag{2.2}$$

For 1/0–coded binary outcome, we can define the causal estimands associated with the population $\mathcal{P}$ and the super-population, respectively, using the same fashion as we define (2.1) and (2.2) for continuous outcome, knowing that $\mathbb{P}(Y^{z=j} = 1) = \mathbb{E}(Y^{z=j})$ if $Y^{z=j}$ is 1/0–coded, $j = 1, 0$.

To conduct inference, there are two approaches: Fisher's approach and Neyman's approach (Imbens and Rubin 2015). To understand the difference between these two approaches, we need to know some history of the idea of potential outcomes.

According to Imbens and Rubin (2015), Neyman proposed the idea of potential outcomes in his Master's thesis in 1923, Fisher applied it to RCTs around the same time, and Rubin in 1974 extended it into a general framework for causal inference in both interventional and non-interventional settings. Imbens and Rubin (2015) also reviewed two theories of statistical inference: Fisher's approach and Neyman's approach.

Fisher proposed an exact test for the following sharp null hypothesis,

$$H_0 : Y_i^{z=1} = Y_i^{z=0}, \quad i = 1, \ldots, N.$$

Neyman proposed the idea of a confidence interval estimator for the population average treatment effect and considered the following pair of hypotheses,

$$H_0 : \theta(\mathcal{P}) = 0 \ \text{ vs. } \ H_a : \theta(\mathcal{P}) \neq 0.$$

Neyman's approach can be generalized easily to construct a confidence interval estimator for θ^* and test the following pair of hypotheses,

$$H_0 : \theta^* = 0 \ \text{ vs. } \ H_a : \theta^* \neq 0.$$

2.2.2 Statistical estimand

"God does not play dice with the universe"—a quote by Albert Einstein, while in RCTs, statisticians play dice with the potential world.

In the real world, of the two potential outcomes $(Y_i^{z=1}, Y_i^{z=0})$, only one is observed. The investigators "determine" which of the two to be observed—by "playing dice"—by assigning treatment 1 to N_1 subjects and treatment 0 to the remaining $N_0 = N - N_1$ subjects. By complete randomization, each version of the treatment assignment is with probability

$$\mathbb{P}(Z_1 = z_1, \ldots, Z_N = z_N) = \binom{N}{N_1}^{-1}, \text{ where } \sum_{i=1}^{N} z_i = N_1.$$

In the above complete randomization, the randomization ratio, $r = N_1/N_0$, is pre-specified. For example, if the ratio is specified as 1:1, then $N_1 = N_0$; if the ratio is 2:1, then $N_1 = 2N_0$.

To link the two worlds—the potential world and the real world, we assume the following **consistency assumption**,

$$Y_i = Y_i^{z=Z_i}, i = 1, \ldots, N. \tag{2.3}$$

In an RCT, the consistency assumption holds if the randomization process is implemented without any operational error.

Therefore, in the real world, we observe $\mathcal{O} = \{(Z_i, Y_i), i = 1, \ldots, N\}$, where Z_i is the treatment assignment and Y_i is the observed one of the two potential outcomes. Let $P_{\mathcal{O}}$ be the underlying distribution that generates observed data $\mathcal{O} = \{(Z_i, Y_i), i = 1, \ldots, N\}$.

Note that, under the complete randomization, we have

$$\mathbb{E}(Z_i) = N_1/N, \ i = 1, \ldots, N, \tag{2.4}$$

$$\mathbb{E}(Z_i Z_j) = N_1(N_1 - 1)/[N(N-1)], \ i \neq j, \tag{2.5}$$

$$\mathrm{Cov}(Z_i, Z_j) = \mathbb{E}(Z_i Z_j) - \mathbb{E}(Z_i)\mathbb{E}(Z_j) = -\frac{N_1 N_0}{N^2(N-1)}, \ i \neq j. \tag{2.6}$$

Thus, $Z_i, i = 1, \ldots, N$, are identically distributed with variable Z (but not independent).

Now we are ready to consider the task of identification, translating the causal estimand defined in terms of P_∞ to a statistical estimand defined in terms of $P_{\mathcal{O}}$. For continuous or binary outcomes, the estimand of interest is

$$\begin{aligned}
\theta^* &= \mathbb{E}(Y^{z=1} - Y^{z=0}) \\
&\overset{(a)}{=} \frac{1}{N}\sum_{i=1}^{N} \mathbb{E}(Y_i^{z=1} - Y_i^{z=0}) \\
&\overset{(b)}{=} \frac{1}{N}\sum_{i=1}^{N} \{\mathbb{E}(Y_i^{z=1}|Z_i = 1) - \mathbb{E}(Y_i^{z=0}|Z_i = 0)\} \\
&\overset{(c)}{=} \frac{1}{N}\sum_{i=1}^{N} \{\mathbb{E}(Y_i|Z_i = 1) - \mathbb{E}(Y_i|Z_i = 0)\} \\
&\overset{(d)}{=} \frac{1}{N}\sum_{i=1}^{N} \mathbb{E}(Y_i|Z_i = 1) - \frac{1}{N}\sum_{i=1}^{N} \mathbb{E}(Y_i|Z_i = 0) \triangleq \theta(P_{\mathcal{O}}),
\end{aligned}$$

where (a) holds because $(Y_i^{z=1}, Y_i^{z=0})$'s are i.i.d. with $(Y^{z=1}, Y^{z=0})$, (b) is ensured by the randomization, (c) holds under the consistency assumption, and (d) holds using the associative property of addition.

To distinguish the two estimands, we refer to θ^* as the **causal estimand** and refer to $\theta(P_{\mathcal{O}})$ as the **statistical estimand**.

Using the finding that, implied by SUTVA and the consistency assumption, $Y_i|(Z_i = 1)$'s are i.i.d. with $Y|(Z = 1)$ and $Y_i|(Z_i = 0)$'s are i.i.d. with $Y|(Z = 0)$, we can rewrite the above statistical estimand as

$$\theta(P_{\mathcal{O}}) = \mathbb{E}(Y|Z = 1) - \mathbb{E}(Y|Z = 0). \tag{2.7}$$

The proof of the equivalence between θ^* and $\theta(P_{\mathcal{O}})$ under the consistency assumption is crucial for three reasons. First, statistical estimand $\theta(P_{\mathcal{O}})$ is estimable in the real world. In the next section, we will construct an estimator to estimate it. Second, via the proof, we make the causal assumptions explicitly. Third, we show that the estimation of $\theta(P_{\mathcal{O}})$ itself is statistically meaningful even if the consistency assumption doesn't hold, but it is a plus to ensure that the estimation of $\theta(P_{\mathcal{O}})$ is also causally meaningful if the consistency assumption is believed to hold.

2.3 Estimator

We consider the construction of an estimator to estimate statistical estimand $\theta(P_{\mathcal{O}})$. The same estimator will estimate the causal estimand θ^* under the consistency assumption because we have just proved the equivalence between θ^* and $\theta(P_{\mathcal{O}})$ under the consistency assumption.

It is straightforward to use sample mean to estimate population mean. Thus, we consider the following estimator to estimate $\theta(P_{\mathcal{O}})$ defined in (2.7),

$$\widehat{\theta}(\mathcal{O}) = \frac{\sum_{i=1}^{N} Z_i Y_i}{N_1} - \frac{\sum_{i=1}^{N} (1 - Z_i) Y_i}{N_0}. \tag{2.8}$$

2.3.1 Expectation of the estimator

We will show that $\widehat{\theta}(\mathcal{O})$ is an unbiased estimator for causal estimands $\theta(\mathcal{P})$ and θ^* under the consistency assumption and for statistical estimand $\theta(P_{\mathcal{O}})$ regardless of whether the consistency assumption holds or not.

When $\theta(\mathcal{P})$ is the target estimand

We can verify that $\widehat{\theta}(\mathcal{O})$ is an unbiased estimator of $\theta(\mathcal{P})$ given $\mathcal{P} = \{(Y_i^{z=1}, Y_i^{z=0}), i = 1, \ldots, N\}$ under the consistency assumption. In fact,

$$\begin{aligned}
\mathbb{E}[\widehat{\theta}(\mathcal{O})|\mathcal{P}] &\overset{(a)}{=} \frac{1}{N_1} \sum_{i=1}^{N} \mathbb{E}[Z_i Y_i|\mathcal{P}] - \frac{1}{N_0} \sum_{i=1}^{N} \mathbb{E}[(1 - Z_i) Y_i|\mathcal{P}] \\
&\overset{(b)}{=} \frac{1}{N_1} \sum_{i=1}^{N} \mathbb{E}[Z_i Y_i^{z=1}|\mathcal{P}] - \frac{1}{N_0} \sum_{i=1}^{N} \mathbb{E}[(1 - Z_i) Y_i^{z=0}|\mathcal{P}] \\
&\overset{(c)}{=} \frac{1}{N_1} \sum_{i=1}^{N} Y_i^{z=1} \mathbb{E}[Z_i|\mathcal{P}] - \frac{1}{N_0} \sum_{i=1}^{N} Y_i^{z=0} \mathbb{E}[(1 - Z_i)|\mathcal{P}] \\
&\overset{(d)}{=} \frac{1}{N_1} \sum_{i=1}^{N} Y_i^{z=1} \frac{N_1}{N} - \frac{1}{N_0} \sum_{i=1}^{N} Y_i^{z=0} \frac{N_0}{N} \\
&\overset{(e)}{=} \frac{1}{N} \sum_{i=1}^{N} Y_i^{z=1} - \frac{1}{N} \sum_{i=1}^{N} Y_i^{z=0} = \theta(\mathcal{P}),
\end{aligned}$$

where (a) holds under the exchange between expectation and summation, (b) holds under the consistency assumption, (c) holds because $Y_i^{z=j}$ is constant conditional on $\mathcal{P}$, (d) holds using (2.4), and (e) holds after canceling N_1 in one term and canceling N_0 in the other term.

When θ^* is the target estimand

Furthermore, we can verify that $\widehat{\theta}(\mathcal{O})$ is also an unbiased estimator of θ^* under the consistency assumption. In fact,

$$\begin{aligned}\mathbb{E}\{\widehat{\theta}(\mathcal{O})\} &\overset{(a)}{=} \mathbb{E}\{\mathbb{E}[\widehat{\theta}(\mathcal{O})|\mathcal{P}]\} \\ &\overset{(b)}{=} \mathbb{E}\left\{\frac{1}{N}\sum_{i=1}^{N} Y_i^{z=1} - \frac{1}{N}\sum_{i=1}^{N} Y_i^{z=0}\right\} \\ &\overset{(c)}{=} \mathbb{E}(Y^{z=1} - Y^{z=0}) = \theta^*,\end{aligned}$$

where (a) holds using the law of iterated expectations, (b) holds using the result that $\mathbb{E}[\widehat{\theta}(\mathcal{O})|\mathcal{P}] = \theta(\mathcal{P})$ under the consistency assumption that is just proved, and (c) holds because $(Y_i^{z=1}, Y_i^{z=0})$'s are i.i.d.$\sim$ $(Y^{z=1}, Y^{z=0})$.

When statistical estimand $\theta(P_{\mathcal{O}})$ is the target estimand

We have proved that $\widehat{\theta}(\mathcal{O})$ is an unbiased estimator of θ^* under the consistency assumption. Recalling that $\theta^* = \theta(P_{\mathcal{O}})$ under the consistency assumption, we can conclude that $\widehat{\theta}(\mathcal{O})$ is an unbiased estimator of $\theta(P_{\mathcal{O}})$.

Actually, we can show that $\widehat{\theta}(\mathcal{O})$ is an unbiased estimator of $\theta(P_{\mathcal{O}})$ without the consistency assumption, which is untestable. Instead, we only need some assumptions on the observed data; that is, $Y_i|(Z_i = 1)$'s are i.i.d. with $Y|(Z = 1)$ and $Y_i|(Z_i = 0)$'s are i.i.d. with $Y|(Z = 0)$. In fact,

$$\begin{aligned}\mathbb{E}\{\widehat{\theta}(\mathcal{O})\} &\overset{(a)}{=} \frac{1}{N_1}\sum_{i=1}^{N}\mathbb{E}[Z_iY_i] - \frac{1}{N_0}\sum_{i=1}^{N}\mathbb{E}[(1-Z_i)Y_i] \\ &\overset{(b)}{=} \frac{N}{N_1}\mathbb{E}[ZY] - \frac{N}{N_0}\sum_{i=1}^{N}\mathbb{E}[(1-Z)Y] \\ &\overset{(c)}{=} \frac{N}{N_1}\mathbb{E}\{\mathbb{E}[ZY|Z]\} - \frac{N}{N_0}\mathbb{E}\{E[(1-Z)Y|Z]\} \\ &\overset{(d)}{=} \frac{N}{N_1}\{\mathbb{E}[ZY|Z=1]\mathbb{P}(Z=1) + 0\times\mathbb{P}(Z=0)\} \\ &\quad - \frac{N}{N_0}\{0\times\mathbb{P}(Z=1) + \mathbb{E}[(1-Z)Y|Z=0]\mathbb{P}(Z=0)\} \\ &\overset{(e)}{=} \frac{N}{N_1}\mathbb{E}[Y|Z=1]\mathbb{P}(Z=1) - \frac{N}{N_0}\mathbb{E}[Y|Z=0]\mathbb{P}(Z=0) \\ &\overset{(f)}{=} \mathbb{E}[Y|Z=1] - \mathbb{E}[Y|Z=0] = \theta(P_{\mathcal{O}}),\end{aligned}$$

where (a) holds after exchanging between expectation and summation, (b) holds because (Z_i, Y_i)'s are identically distributed with (Z, Y), (c) holds using the law of iterated expectations, (d) holds using the definition of conditional expectation, (e) holds after removing zero terms, and (f) holds using (2.4).

2.3.2 Variance of the estimator

In order to conduct statistical inferences based on $\widehat{\theta}(\mathcal{O})$, we need to have an estimator for the variance of $\widehat{\theta}(\mathcal{O})$. We have shown that $\widehat{\theta}(\mathcal{O})$ is an unbiased estimator for $\theta(\mathcal{P})$, θ^*, and $\theta(P_{\mathcal{O}})$, respectively. Therefore, we need to develop the corresponding formula to calculate the variance of $\widehat{\theta}(\mathcal{O})$ when the target estimand is $\theta(\mathcal{P})$, θ^*, or $\theta(P_{\mathcal{O}})$.

When $\theta(\mathcal{P})$ is the target estimand

When $\theta(\mathcal{P})$ is the target estimand, we need to find a formula for $\mathbb{V}[\widehat{\theta}(\mathcal{O})|\mathcal{P}]$, the variance of $\widehat{\theta}(\mathcal{O})$ given $\mathcal{P} = \{(Y_i^{z=1}, Y_i^{z=0}), i = 1, \dots, N\}$. For this aim, under the consistency assumption, we rewrite $\widehat{\theta}(\mathcal{O})$ in terms of potential outcomes,

$$
\begin{aligned}
\widehat{\theta}(\mathcal{O}) =& \frac{\sum_{i=1}^N Z_i Y_i^{z=1}}{N_1} - \frac{\sum_{i=1}^N (1 - Z_i) Y_i^{z=0}}{N_0} \\
=& \sum_{i=1}^N Z_i \left(\frac{Y_i^{z=1}}{N_1} + \frac{Y_i^{z=0}}{N_0} \right) - \sum_{i=1}^N \frac{Y_i^{z=0}}{N_0} \\
=& \sum_{i=1}^N c_i Z_i - \sum_{i=1}^N \frac{Y_i^{z=0}}{N_0},
\end{aligned}
$$

where $c_i = Y_i^{z=1}/N_1 + Y_i^{z=0}/N_0$. Then we have

$$
\begin{aligned}
\mathbb{V}[\widehat{\theta}(\mathcal{O})|\mathcal{P}] =& \mathbb{V}\left(\sum_{i=1}^N c_i Z_i \middle| \mathcal{P} \right) \\
=& \mathbb{E}\left[\left(\sum_{i=1}^N c_i Z_i \right)^2 \middle| \mathcal{P} \right] - \left[\mathbb{E}\left(\sum_{i=1}^N c_i Z_i \middle| \mathcal{P} \right) \right]^2 \\
=& \sum_{i=1}^N c_i^2 \mathbb{E}(Z_i^2) + 2 \sum_{i<j} c_i c_j \mathbb{E}(Z_i Z_j) - \left[\sum_{i=1}^N c_i \mathbb{E}(Z_i) \right]^2 \\
=& \frac{N_1}{N} \sum_{i=1}^N c_i^2 + 2 \frac{N_1(N_1 - 1)}{N(N-1)} \sum_{i<j} c_i c_j - \left[\sum_{i=1}^N c_i \frac{N_1}{N} \right]^2 \\
=& \frac{N_1 N_0}{N^2} \sum_{i=1}^N c_i^2 - \frac{2 N_1 N_0}{N^2(N-1)} \sum_{i<j} c_i c_j. \qquad (2.9)
\end{aligned}
$$

When θ^* is the target estimand

When θ^* is the target estimand, we need to find a formula for $\mathbb{V}[\widehat{\theta}(\mathcal{O})]$, the unconditional variance of $\widehat{\theta}(\mathcal{O})$ provided that $(Y_i^{z=1}, Y_i^{z=0}), i = 1, \ldots, N$, are i.i.d. with $(Y^{z=1}, Y^{z=0})$. For this aim, we consider the following decomposition,

$$\mathbb{V}\{\widehat{\theta}(\mathcal{O})\} = \mathbb{E}\{\mathbb{V}[\widehat{\theta}(\mathcal{O})|\mathcal{P}]\} + \mathbb{V}\{\mathbb{E}[\widehat{\theta}(\mathcal{O})|\mathcal{P}]\}. \tag{2.10}$$

Note that the first term on the right-hand-side (RHS) of (2.10) is the expectation of $\mathbb{V}[\widehat{\theta}(\mathcal{O})|\mathcal{P}]$ that is derived in (2.9). Thus,

$$\mathbb{E}\{\mathbb{V}[\widehat{\theta}(\mathcal{O})|\mathcal{P}]\} = \frac{N_1 N_0}{N^2} \sum_{i=1}^{N} \mathbb{E}(c_i^2) - \frac{2N_1 N_0}{N^2(N-1)} \sum_{i<j} \mathbb{E}(c_i c_j).$$

Noting that

$$\begin{aligned} c_i^2 &= \left(\frac{Y_i^{z=1}}{N_1}\right)^2 + \left(\frac{Y_i^{z=0}}{N_0}\right)^2 + 2\left(\frac{Y_i^{z=1}}{N_1}\right)\left(\frac{Y_i^{z=0}}{N_0}\right), \\ c_i c_j &= \left(\frac{Y_i^{z=1}}{N_1} + \frac{Y_i^{z=0}}{N_0}\right)\left(\frac{Y_j^{z=1}}{N_1} + \frac{Y_j^{z=0}}{N_0}\right), \quad i < j, \end{aligned}$$

we have

$$\begin{aligned} \mathbb{E}(c_i^2) &= \frac{\mathbb{E}[(Y^{z=1})^2]}{N_1^2} + \frac{\mathbb{E}[(Y^{z=0})^2]}{N_0^2} + \frac{2\mathbb{E}(Y^{z=1}Y^{z=0})}{N_1 N_0}, \\ \mathbb{E}(c_i c_j) &= \left[\frac{\mathbb{E}(Y^{z=1})}{N_1} + \frac{\mathbb{E}(Y^{z=0})}{N_0}\right]^2, \quad i < j. \end{aligned}$$

Thus, $\mathbb{E}\{\mathbb{V}[\widehat{\theta}(\mathcal{O})|\mathcal{P}]\}$ is equal to

$$\frac{N_1 N_0}{N}\left[\frac{\mathbb{E}(Y^{z=1})^2}{N_1^2} + \frac{\mathbb{E}(Y^{z=0})^2}{N_0^2} + \frac{2\mathbb{E}(Y^{z=1}Y^{z=0})}{N_1 N_0} - \left[\frac{\mathbb{E}(Y^{z=1})}{N_1} + \frac{\mathbb{E}(Y^{z=0})}{N_0}\right]^2\right].$$

Now consider the second term on RHS of (2.10). Since

$$\mathbb{E}[\widehat{\theta}(\mathcal{O})|\mathcal{P}] = \frac{1}{N}\sum_{i=1}^{N}(Y_i^{z=1} - Y_i^{z=0}),$$

we have

$$\begin{aligned} \mathbb{V}\{\mathbb{E}[\widehat{\theta}(\mathcal{O})|\mathcal{P}]\} &= \frac{1}{N}\mathbb{V}(Y^{z=1} - Y^{z=0}) \\ &= \frac{\mathbb{E}[(Y^{z=1} - Y^{z=0})^2]}{N} - \frac{[\mathbb{E}(Y^{z=1}) - \mathbb{E}(Y^{z=0})]^2}{N}. \end{aligned}$$

Using some elementary algebra, we obtain

$$\mathbb{V}\{\widehat{\theta}(\mathcal{O})\} = \mathbb{E}\{\mathbb{V}[\widehat{\theta}(\mathcal{O})|\mathcal{P}]\} + \mathbb{V}\{\mathbb{E}[\widehat{\theta}(\mathcal{O})|\mathcal{P}]\} = \frac{\mathbb{V}(Y^{z=1})}{N_1} + \frac{\mathbb{V}(Y^{z=0})}{N_0}. \quad (2.11)$$

The above processes of deriving $\mathbb{V}[\widehat{\theta}(\mathcal{O})|\mathcal{P}]$ and $\mathbb{V}\{\widehat{\theta}(\mathcal{O})\}$ are motivated by the corresponding ones provided in Appendix B of Chapter 6 of Imbens and Rubin (2015). The derivation is tedious, you may omit it without sacrificing any key point. Fortunately, we have a much easier method for deriving the formula (2.11) for $\mathbb{V}[\widehat{\theta}(\mathcal{O})]$ using the following decomposition,

$$\mathbb{V}\{\widehat{\theta}(\mathcal{O})\} = \mathbb{E}\{\mathbb{V}[\widehat{\theta}(\mathcal{O})|\mathcal{Z}]\} + \mathbb{V}\{\mathbb{E}[\widehat{\theta}(\mathcal{O})|\mathcal{Z}]\}, \quad (2.12)$$

where $\mathcal{Z} = \{Z_1, \ldots, Z_N\}$ is a given realization of the randomization. In fact, the first term on RHS of (2.12) is

$$\begin{aligned}
&\mathbb{E}\{\mathbb{V}[\widehat{\theta}(\mathcal{O})|\mathcal{Z}]\} \\
=&\mathbb{E}\left\{\mathbb{V}\left[\frac{\sum_{i=1}^N Z_i Y_i^{z=1}}{N_1} - \frac{\sum_{i=1}^N (1-Z_i) Y_i^{z=0}}{N_0}\middle|\mathcal{Z}\right]\right\} \\
=&\mathbb{E}\left\{\frac{\mathbb{V}(Y^{z=1})}{N_1} + \frac{\mathbb{V}(Y^{z=0})}{N_0}\right\} = \frac{\mathbb{V}(Y^{z=1})}{N_1} + \frac{\mathbb{V}(Y^{z=0})}{N_0}.
\end{aligned}$$

And the second item on RHS of (2.12) is

$$\begin{aligned}
&\mathbb{V}\{\mathbb{E}[\widehat{\theta}(\mathcal{O})|\mathcal{Z}]\} \\
=&\mathbb{V}\left\{\mathbb{E}\left[\frac{\sum_{i=1}^N Z_i Y_i^{z=1}}{N_1} - \frac{\sum_{i=1}^N (1-Z_i) Y_i^{z=0}}{N_0}\middle|\mathcal{Z}\right]\right\} \\
=&\mathbb{V}\left\{\mathbb{E}[Y^{z=1}] - \mathbb{E}[Y^{z=0}]\right\} = 0.
\end{aligned}$$

Combining these two terms, we derive the same formula (2.11).

When $\theta(P_{\mathcal{O}})$ is the target estimand

When $\theta(P_{\mathcal{O}})$ is the target estimand, we don't need the consistency assumption. Instead, we only need some assumption on the observed data; that is, $Y_i|(Z_i = 1), i = 1, \ldots, N$, are i.i.d. with $Y|(Z = 1)$, $Y_i|(Z_i = 0), i = 1, \ldots, N$, are i.i.d. with $Y|(Z = 0)$, and $Y_i|(Z_i = z_i), i = 1, \ldots, N$, are independent. this realistic assumption, we can obtain a formula for the variance of $\widehat{\theta}(\mathcal{O})$ using the same decomposition in (2.12).

In fact, the first item on RHS of (2.12) is

$$\begin{aligned}
\mathbb{E}\{\mathbb{V}[\widehat{\theta}(\mathcal{O})|\mathcal{Z}]\} &= \mathbb{E}\left\{\mathbb{V}\left[\frac{\sum_{i=1}^{N} Z_iY_i}{N_1} - \frac{\sum_{i=1}^{N}(1-Z_i)Y_i}{N_0}\middle|\mathcal{Z}\right]\right\} \\
=&\mathbb{E}\left\{\mathbb{V}\left[\frac{\sum_{\{i:Z_i=1\}} Y_i}{N_1} - \frac{\sum_{\{i:Z_i=0\}} Y_i}{N_0}\middle|\mathcal{Z}\right]\right\} \\
=&\mathbb{E}\left\{\mathbb{V}\left[\frac{\sum_{\{i:Z_i=1\}} Y_i}{N_1}\middle|\mathcal{Z}\right] + \mathbb{V}\left[\frac{\sum_{\{i:Z_i=0\}} Y_i}{N_0}\middle|\mathcal{Z}\right]\right\} \\
=&\mathbb{E}\left\{\frac{\mathbb{V}(Y|Z=1)}{N_1} + \frac{\mathbb{V}(Y|Z=0)}{N_0}\right\} \\
=&\frac{\mathbb{V}(Y|Z=1)}{N_1} + \frac{\mathbb{V}(Y|Z=0)}{N_0}.
\end{aligned}$$

And the second item on RHS of (2.12) is

$$\begin{aligned}
\mathbb{V}\{\mathbb{E}[\widehat{\theta}(\mathcal{O})|\mathcal{Z}]\} &= \mathbb{V}\left\{\mathbb{E}\left[\frac{\sum_{i=1}^{N} Z_iY_i}{N_1} - \frac{\sum_{i=1}^{N}(1-Z_i)Y_i}{N_0}\middle|\mathcal{Z}\right]\right\} \\
=&\mathbb{V}\left\{\mathbb{E}[Y|Z=1] - \mathbb{E}[Y|Z=0]\right\} = 0.
\end{aligned}$$

Combining the above two terms, we have

$$\mathbb{V}\{\widehat{\theta}(\mathcal{O})\} = \frac{\mathbb{V}(Y|Z=1)}{N_1} + \frac{\mathbb{V}(Y|Z=0)}{N_0}. \tag{2.13}$$

2.3.3 Statistical inference

In order to conduct statistical inference for statistical estimand (2.7), we consider the estimator proposed in (2.8), along with its variance developed in (2.13). Therefore, we need to obtain an estimator for (2.13). For this aim, we rearrange the data into two arms, $\{Y_{1i}, i = 1, \ldots, N_1\}$ that form the treatment arm and $\{Y_{0i}, i = 1, \ldots, N_0\}$ that form the control arm. Using the arm-specific data, we can estimate $\mathbb{V}(Y|Z = j)$ using its sample variance,

$$\widehat{\sigma}_j^2 = \frac{1}{N_j - 1}\sum_{i=1}^{N_j}\left(Y_{ji} - \frac{1}{N_j}\sum_{i=1}^{N_j} Y_{ji}\right)^2, \quad j = 1, 0.$$

Then we obtain an estimator for the standard error (SE) of $\widehat{\theta}(\mathcal{O})$,

$$\widehat{\mathrm{SE}} = \left(\frac{\widehat{\sigma}_1^2}{N_1} + \frac{\widehat{\sigma}_0^2}{N_0}\right)^{1/2}.$$

Thus, we can construct a 95% confidence interval (CI) estimator for $\theta(P_{\mathcal{O}})$,

$$\text{CI} = \widehat{\theta}(\mathcal{O}) \pm 1.96 \times \widehat{\text{SE}}.$$

In addition, using the Z-test reviewed in Chapter 1, we can obtain the p-value for hypotheses $H_0 : \theta(P_{\mathcal{O}}) = 0$ vs. $H_a : \theta(P_{\mathcal{O}}) \neq 0$. By obtaining CI and p-value, we claim that we are conducting statistical inference.

Moreover, because $\theta(P_{\mathcal{O}}) = \theta^*$ under the consistency assumption, the above CI is also a confidence interval estimator for θ^* and the above p-value is also for hypotheses $H_0 : \theta^* = 0$ vs. $H_a : \theta^* \neq 0$. With this reasoning, we claim we are conducting causal inference beyond statistical inference.

2.4 Common Types of Randomization

2.4.1 Simple randomization

In the above complete randomization with ratio r:1, we enroll N subjects all together, and then randomize $N_1 = Nr/(1 + r)$ subjects to treatment 1 and the remaining $N_0 = N/(1 + r)$ subjects to treatment 0. This randomization procedure is applicable to populations with small size.

For a population with, moderately large or large size, a more realistic randomization procedure is simple randomization. That is, we enroll subjects one by one, and for the ith subject we enroll in, we "flip a coin" to randomize the subject to treatment 1 with probability $\mathbb{P}(Z_i = 1) = r/(1 + r) = g_1$ and to treatment 0 with probability $\mathbb{P}(Z_i = 0) = 1/(1 + r) = g_0$.

The difference between simple randomization and the complete randomization discussed earlier is that in simple randomization Z_i, $i = 1, \dots, N$, are i.i.d. with Z where $\mathbb{P}(Z = 1) = g_1$ and $\mathbb{P}(Z = 0) = g_0$.

Causal estimand

The randomization process won't impact the potential world; instead, it only impacts which version of the two potential outcomes is observed in the real world. Therefore, in the potential world, that $(Y_i^{z=1}, Y_i^{z=1}), i = 1, \dots, N$, are i.i.d. with $(Y^{z=1}, Y^{z=0})$ still holds and how to define the causal estimand of interest remains the same.

Statistical estimand

Now let's consider the task of identification under the consistency assumption. Throughout the book, we will see that, for almost all the causal estimands, there are two basic strategies to complete the task of identification, translating the causal estimand into a statistical estimand.

The two basic strategies will be referred to as the standardization strategy and the weighting strategy (Fang 2020; Hernán and Robins 2020), respectively. We have seen the standardization strategy in the previous section. By the standardization strategy, we consider the following identification process,

$$\begin{aligned}\theta^* &= \mathbb{E}(Y^{z=1}) - \mathbb{E}(Y^{z=0}) \\ &\overset{(a)}{=} \mathbb{E}[Y^{z=1}|Z=1] - \mathbb{E}[Y^{z=0}|Z=0] \\ &\overset{(b)}{=} \mathbb{E}[Y|Z=1] - \mathbb{E}[Y|Z=0],\end{aligned}$$

where (a) holds because of randomization and (b) holds because of the consistency assumption.

By the weighting strategy, we consider the following identification process,

$$\begin{aligned}\theta^* &= \mathbb{E}(Y^{z=1}) - \mathbb{E}(Y^{z=0}) \\ &\overset{(a)}{=} \mathbb{E}\left[\frac{I(Z=1)}{\mathbb{P}(Z=1)}Y^{z=1}\right] - \mathbb{E}\left[\frac{I(Z=0)}{\mathbb{P}(Z=0)}Y^{z=0}\right] \\ &\overset{(b)}{=} \mathbb{E}\left[\frac{I(Z=1)Y}{\mathbb{P}(Z=1)}\right] - \mathbb{E}\left[\frac{I(Z=0)Y}{\mathbb{P}(Z=0)}\right],\end{aligned}$$

where (a) holds because by simple randomization we have

$$\mathbb{E}\left[\frac{I(Z=1)}{\mathbb{P}(Z=1)}Y^{z=1}\right] = \mathbb{E}\left[\frac{I(Z=j)}{\mathbb{P}(Z=j)}\right]\mathbb{E}\left[Y^{z=j}\right] = \mathbb{E}\left[Y^{z=j}\right],$$

and (b) holds because of the consistency assumption. Therefore, under the consistency assumption, causal estimand θ^* is equal to the following statistical estimand,

$$\begin{aligned}\theta(P_{\mathcal{O}}) &= \mathbb{E}\left[\frac{I(Z=1)Y}{\mathbb{P}(Z=1)}\right] - \mathbb{E}\left[\frac{I(Z=0)Y}{\mathbb{P}(Z=0)}\right] \\ &= \mathbb{E}\left[\frac{ZY}{\mathbb{P}(Z=1)}\right] - \mathbb{E}\left[\frac{(1-Z)Y}{\mathbb{P}(Z=0)}\right] \\ &= \mathbb{E}\left[\frac{ZY}{g_1}\right] - \mathbb{E}\left[\frac{(1-Z)Y}{1-g_1}\right].\end{aligned} \tag{2.14}$$

Estimator, along with its expectation, variance, and inference

To estimate statistical estimand (2.14), using the sample mean to replace the population mean, we construct the following estimator,

$$\widehat{\theta}(\mathcal{O}) = \frac{\sum_{i=1}^N Z_i Y_i}{N g_1} - \frac{\sum_{i=1}^N (1-Z_i) Y_i}{N(1-g_1)}. \tag{2.15}$$

We can verify that this estimator is an unbiased estimator for $\theta(P_{\mathcal{O}})$ under the realistic assumption that (Z_i, Y_i)'s are i.i.d. with (Z, Y). In fact,

$$\begin{aligned}
\mathbb{E}\{\widehat{\theta}(\mathcal{O})\} &\overset{(a)}{=} \mathbb{E}\{\mathbb{E}[\widehat{\theta}(\mathcal{O})|\mathcal{Z}]\} \\
&\overset{(b)}{=} \mathbb{E}\left\{\mathbb{E}\left[\frac{\sum_{i=1}^N Z_i Y_i}{N g_1} - \frac{\sum_{i=1}^N (1-Z_i) Y_i}{N(1-g_1)}\middle| \mathcal{Z}\right]\right\} \\
&\overset{(c)}{=} \mathbb{E}\left\{\frac{\sum_{i=1}^N Z_i \mathbb{E}(Y_i|Z_i=1)}{N g_1} - \frac{\sum_{i=1}^N (1-Z_i)\mathbb{E}(Y_i|Z_i=0)}{N(1-g_1)}\right\} \\
&\overset{(d)}{=} \mathbb{E}\left\{\frac{\sum_{i=1}^N Z_i \mathbb{E}(Y|Z=1)}{N g_1} - \frac{\sum_{i=1}^N (1-Z_i)\mathbb{E}(Y|Z=0)}{N(1-g_1)}\right\} \\
&\overset{(e)}{=} \mathbb{E}(Y|Z=1) - \mathbb{E}(Y|Z=0) = \theta(P_{\mathcal{O}}),
\end{aligned}$$

where (a) uses the law of iterated expectations, (b) uses (2.15), (c) holds because Z_i is constant conditional on Z_i, (d) holds because (Y_i, Z_i)'s are i.i.d., and (e) holds because $\mathbb{E}(Z_i) = g_1$.

We can also develop a formula for its variance. In fact,

$$\begin{aligned}
\mathbb{V}\{\widehat{\theta}(\mathcal{O})\} &\overset{(a)}{=} \mathbb{E}\{\mathbb{V}[\widehat{\theta}(\mathcal{O})|\mathcal{Z}]\} + \mathbb{V}\{\mathbb{E}[\widehat{\theta}(\mathcal{O})|\mathcal{Z}]\} \\
&\overset{(b)}{=} \mathbb{E}\left\{\mathbb{V}\left[\frac{\sum_{i=1}^N Z_i Y_i}{N g_1} - \frac{\sum_{i=1}^N (1-Z_i) Y_i}{N(1-g_1)}\middle| \mathcal{Z}\right]\right\} + 0 \\
&\overset{(c)}{=} \mathbb{E}\left\{\mathbb{V}\left[\frac{\sum_{i=1}^N Z_i Y_i}{N g_1}\middle| \mathcal{Z}\right]\right\} + \mathbb{E}\left\{\mathbb{V}\left[\frac{\sum_{i=1}^N (1-Z_i) Y_i}{N(1-g_1)}\middle| \mathcal{Z}\right]\right\}. \\
&\overset{(d)}{=} \mathbb{E}\left\{\sum_{i=1}^N \frac{Z_i^2 \mathbb{V}(Y_i|Z_i=1)}{(N g_1)^2}\right\} + \mathbb{E}\left\{\sum_{i=1}^N \frac{(1-Z_i)^2 \mathbb{V}(Y_i|Z_i=0)}{[N(1-g_1)]^2}\right\} \\
&\overset{(e)}{=} \mathbb{E}\left\{\sum_{i=1}^N \frac{Z_i \mathbb{V}(Y|Z=1)}{(N g_1)^2}\right\} + \mathbb{E}\left\{\sum_{i=1}^N \frac{(1-Z_i) \mathbb{V}(Y|Z=0)}{[N(1-g_1)]^2}\right\} \\
&\overset{(f)}{=} \frac{\mathbb{V}(Y|Z=1)}{N g_1} + \frac{\mathbb{V}(Y|Z=0)}{N(1-g_1)},
\end{aligned}$$

where (a) uses the decomposition of variance, (b) holds because the second term of the decomposition is the variance of a constant (hence becomes zero), (c) holds because $Z_i(1-Z_i) = 0$, (d) holds because (Z_i, Y_i)'s are independent, (e) holds because (Z_i, Y_i)'s are identically distributed, and (f) holds because $\mathbb{E}(Z_i) = g_1$.

Note that this formula of the variance is almost the same as the one in (2.13) knowing that $N_1 \doteq N g_1$ and $N_0 \doteq N(1-g_1)$.

We rearrange the data into two arms, $\{Y_{1i}, i = 1, \ldots, N_1\}$ that form the treatment arm and $\{Y_{0i}, i = 1, \ldots, N_0\}$ that form the control arm. Using the arm-specific data, we can estimate $\mathbb{V}(Y|Z=j)$ using its sample variance, $\widehat{\sigma}_j^2$, $j=1,0$. Then we can obtain an estimator for the SE of $\widehat{\theta}(\mathcal{O})$,

$$\widehat{\text{SE}} = \left[\frac{\widehat{\sigma}_1^2}{Ng_1} + \frac{\widehat{\sigma}_0^2}{N(1-g_1)}\right]^{1/2}.$$

Thus, we can obtain both 95% CI for $\theta(P_{\mathcal{O}})$ and p-value for $H_0 : \theta(P_{\mathcal{O}}) = 0$ vs. $H_a : \theta(P_{\mathcal{O}}) \neq 0$.

2.4.2 Block randomization

In practice, the randomization is often performed in blocks. For example, if the ratio is 1:1, we may consider blocks of four; that is, we enroll subjects in blocks of four, within each block, two subjects are randomly assigned to treatment 1 and the other two subjects are assigned to treatment 0. As another example, if the ratio is 2:1, we may consider blocks of six; that is, we enroll subjects in blocks of six, within each block, four subjects are randomly assigned to treatment 1 and the other two subjects are assigned to treatment 0.

In general, if we want to enroll $N = nm$ subjects, with n blocks of size m, and the ratio within each block is set as r:1, then by block randomization, we can ensure the ratio of two arms stay close to r:1 during the entire enrollment process.

Block randomization enjoys the strength of both the complete randomization discussed earlier and the simple randomization. Actually, all three types of randomization are complete randomization in the sense that every subject has the same chance to receive either treatment 1 or 0, but here we call the randomization discussed in Sections 2.2–2.3 as the complete randomization to distinguish it from the other two types of randomization. In addition, the complete randomization is one special case of block randomization with only one block and simple randomization is one special case of block randomization with blocks of size one. Therefore, it is straightforward to generalize the results from single randomization to block randomization.

Causal estimand

The randomization process won't impact the potential world, so the causal estimand of interest remains the same. For example, the target estimand is θ^* defined in (2.2).

Statistical estimand

Under the consistency assumption, we can translate the causal estimand θ^* into the same statistical estimand defined in (2.14).

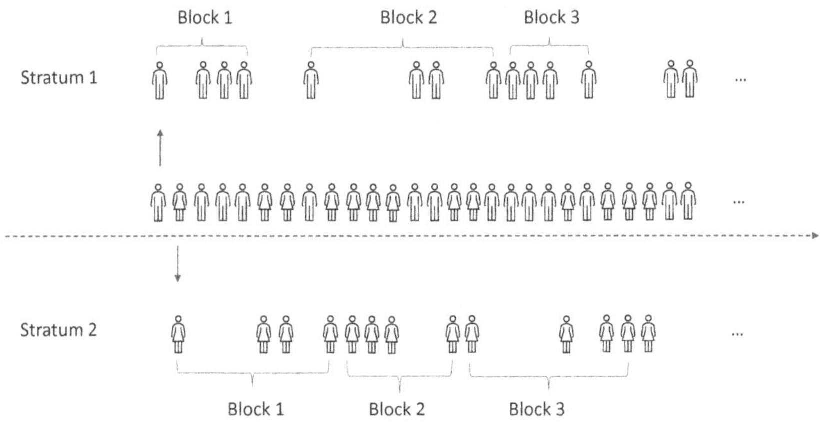

FIGURE 2.2
An illustration of stratified randomization

Estimator, along with its expectation, variance, and inference

Similarly, consider the estimator defined in (2.15). Therefore, we can easily generalize the discussion on its expectation, variance, and inference from simple randomization to block randomization.

2.4.3 Stratified randomization

Let's start with an excerpt from ICH E9 on stratification.

> "More generally, stratification by important prognostic factors measured at baseline (e.g. severity of disease, [...]) may sometimes be valuable in order to promote balanced allocation within strata; this has greater potential benefit in small trials. [...] Factors on which randomisation has been stratified should be accounted for later in the analysis."—ICH E9

In Figure 2.2, patients are enrolled one by one and are stratified according to their gender. Within each stratum, a block randomization is implemented.

To understand why "factors on which randomization has been stratified should be accounted for later in the analysis," we supply some mathematical symbols. Let S be the vector of stratification factors, with K strata, indexed by $S = 1, \ldots, K$. Within stratum $S = k$, we implement block randomization

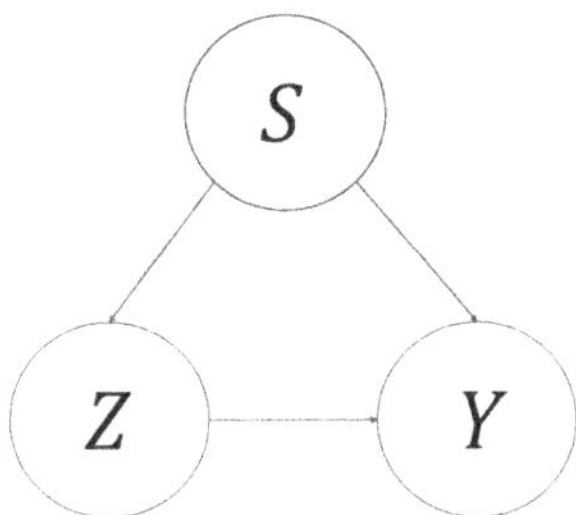

FIGURE 2.3
The DAG of stratified randomization

using ratio r_k:1; that is, the number of subjects assigned to treatment 1 is r_k times the number of those assigned to treatment 0, $k = 1, \ldots, K$.

For example, we consider severity of disease as a stratification factor, with $S = 1$ for severe and $S = 0$ for moderate or mild. In this example, we may consider 1:1 ratio for stratum $S = 0$, while we may consider 2:1 ratio for stratum $S = 1$ to assign more severe subjects to treatment 1.

Figure 2.3 shows the DAG of the stratified randomization, in which the marginal association between between Z and Y is from two paths: causal-effect path $Z \rightarrow Y$ and common-cause path $Z \leftarrow S \rightarrow Y$. Therefore, if we are interested in the causal effect of Z on Y, we need to account for the stratification factor S.

Causal estimand

Start with population $\mathcal{P} = \{(S_i, Y_i^{z=1}, Y_i^{z=0}), i = 1, \ldots, N\}$, where S_i is the stratum status of subject i. Let $N(k) = \#\{i : S_i = k\}$, the number of subjects in stratum k, and $\pi_k = N(k)/N$, the proportion of subjects in stratum k. We arrange the population $\mathcal{P}$ into K sub-populations according to the stratum status, denoted as $\mathcal{P}(k) = \{(Y_{ki}^{z=1}, Y_{ki}^{z=0}), i = 1, \ldots, N(k)\}$, $k = 1, \ldots, K$.

We define the following causal estimand associated with population $\mathcal{P}$,

$$\theta(\mathcal{P}) = \frac{1}{N} \sum_{i=1}^{N} \left(Y_i^{z=1} - Y_i^{z=0}\right) = \frac{1}{N} \sum_{k=1}^{K} \sum_{i=1}^{N(k)} \left(Y_{ki}^{z=1} - Y_{ki}^{z=0}\right). \tag{2.16}$$

Next we conceptually construct a super-population from which those N subjects are drawn such that $(S_i, Y_i^{z=1}, Y_i^{z=0}), i = 1, \ldots, N$, are i.i.d. with $(S, Y^{z=1}, Y^{z=0})$, where S is a random variable with $\mathbb{P}(S = k) = \pi_k, k = 1, \ldots, K$. Thus, we define the following causal estimand associated with the super-population,

$$
\begin{aligned}
\theta^* = \mathbb{E}[\theta(\mathcal{P})] &= \mathbb{E}\left[\frac{1}{N}\sum_{i=1}^{N}\left(Y_i^{z=1} - Y_i^{z=0}\right)\right] \\
&\overset{(a)}{=} \mathbb{E}\left(Y^{z=1} - Y^{z=0}\right) \\
&\overset{(b)}{=} \mathbb{E}\left[\mathbb{E}\left(Y^{z=1} - Y^{z=0} \middle| S\right)\right] \\
&\overset{(c)}{=} \sum_{k=1}^{K}\mathbb{E}\left(Y^{z=1} - Y^{z=0}|S=k\right)\mathbb{P}(S=k) \\
&\overset{(d)}{=} \sum_{k=1}^{K}\pi_k\mathbb{E}\left(Y^{z=1} - Y^{z=0}|S=k\right),
\end{aligned}
\tag{2.17}
$$

where (a) holds because $(Y_i^{z=1}, Y_i^{z=0})$'s are i.i.d. with $(Y^{z=1}, Y^{z=0})$, (b) uses the law of iterated expectations, (c) uses the definition of conditional expectation, and (d) holds because $\mathbb{P}(S=k) = \pi_k$.

Note that we may be interested in other estimands, which are defined using different weights other than π_k's. Or, we may be interested in an estimand associated with sub-population $\mathcal{P}(k)$ only.

Statistical estimand

Recall that within stratum k we implement block randomization using the ratio r_k:1. Then, we observe data $\mathcal{O} = \cup_{k=1}^{K}\mathcal{O}(k)$, where

$$\mathcal{O}(k) = \{(Z_{ki}, Y_{ki}), i = 1, \ldots, N(k)\},$$

where $Z_{ki} = 1$ or 0 is the random assignment for subject i in stratum k.

The consistency assumption assumes that $Y_{ki} = Y_{ki}^{z=Z_{ki}}$, for $k = 1, \ldots, K$ and $i = 1, \ldots, N(k)$. Consequently, $(Z_{ki}, Y_{ki}), i = 1, \ldots, N(k)$, are identically distributed with $(Z_{(k)}, Y_{(k)})$, where $\mathbb{P}(Z_{(k)} = 1) = g_{k1} = r_k/(r_k+1)$.

Let's translate causal estimand θ^* to a statistical estimand:

$$
\begin{aligned}
\theta^* &\overset{(a)}{=} \sum_{k=1}^{K}\frac{N(k)}{N}\mathbb{E}\left(Y^{z=1} - Y^{z=0}|S=k\right) \\
&\overset{(b)}{=} \frac{1}{N}\sum_{k=1}^{K}\sum_{i=1}^{N(k)}\mathbb{E}\left(Y_{ki}^{z=1} - Y_{ki}^{z=0}\right) \\
&\overset{(c)}{=} \frac{1}{N}\sum_{k=1}^{K}\sum_{i=1}^{N(k)}\left[\mathbb{E}(Y_{ki}^{z=1}|Z_{ki}=1) - \mathbb{E}(Y_{ki}^{z=0}|Z_{ki}=0)\right] \\
&\overset{(d)}{=} \frac{1}{N}\sum_{k=1}^{K}\sum_{i=1}^{N(k)}\left[\mathbb{E}(Y_{ki}|Z_{ki}=1) - \mathbb{E}(Y_{ki}|Z_{ki}=0)\right] \\
&\overset{(e)}{=} \sum_{k=1}^{K}\frac{N(k)}{N}\left[\mathbb{E}(Y_{(k)}|Z_{(k)}=1) - \mathbb{E}(Y_{(k)}|Z_{(k)}=0)\right] \triangleq \theta(P_{\mathcal{O}}),
\end{aligned}
$$

where (a) holds because $\pi_k = N(k)/N$, (b) holds because $(Y_{ki}^{z=1}, Y_{ki}^{z=0})$'s are i.i.d. with $(Y^{z=1}, Y^{z=0})|S = k$, (c) holds because the stratified randomization ensures that Z_{ki} is independent of $(Y_{ki}^{z=1}, Y_{ki}^{z=0})$ within stratum k, (d) holds under the consistency assumption, and (e) holds because $Y_{ki}, i = 1, \dots, N(k)$ are identically distributed.

Estimator, along with its expectation, variance, and inference

To estimate the following statistical estimand,

$$\theta(P_{\mathcal{O}}) = \sum_{k=1}^{K} \frac{N(k)}{N} \left[\mathbb{E}(Y_{(k)}|Z_{(k)} = 1) - \mathbb{E}(Y_{(k)}|Z_{(k)} = 0)\right], \tag{2.18}$$

we consider the following estimator,

$$\widehat{\theta}(\mathcal{O}) = \sum_{k=1}^{K} \frac{N(k)}{N} \left[\frac{\sum_{i=1}^{N(k)} Z_{ki}Y_{ki}}{N(k)g_{1k}} - \frac{\sum_{i=1}^{N(k)}(1 - Z_{ki})Y_{ki}}{N(k)(1 - g_{k1})}\right]. \tag{2.19}$$

Since we implement block randomization for stratum k, we have

$$\begin{aligned}&\mathbb{E}\left[\frac{\sum_{i=1}^{N(k)} Z_{ki}Y_{ki}}{N(k)g_{k1}} - \frac{\sum_{i=1}^{N(k)}(1 - Z_{ki})Y_{ki}}{N(k)(1 - g_{k1})}\right]\\ =&\mathbb{E}(Y_{(k)}|Z_{(k)} = 1) - \mathbb{E}(Y_{(k)}|Z_{(k)} = 0)\end{aligned}$$

and

$$\begin{aligned}&\mathbb{V}\left[\frac{\sum_{i=1}^{N(k)} Z_{ki}Y_{ki}}{N(k)g_{1k}} - \frac{\sum_{i=1}^{N(k)}(1 - Z_{ki})Y_{ki}}{N(k)(1 - g_{k1})}\right]\\ =&\frac{\mathbb{V}(Y_{(k)}|Z_{(k)} = 1)}{N(k)g_{k1}} + \frac{\mathbb{V}(Y_{(k)}|Z_{(k)} = 0)}{N(k)(1 - g_{k1})}.\end{aligned}$$

Thus, we show that

$$\mathbb{E}\{\widehat{\theta}(\mathcal{O})\} = \theta(P_{\mathcal{O}}),$$

$$\mathbb{V}\{\widehat{\theta}(\mathcal{O})\} = \sum_{k=1}^{K} \pi_k^2 \left[\frac{\mathbb{V}(Y_{(k)}|Z_{(k)} = 1)}{N(k)g_{k1}} + \frac{\mathbb{V}(Y_{(k)}|Z_{(k)} = 0)}{N(k)(1 - g_{k1})}\right].$$

We can estimate $\mathbb{V}(Y_{(k)}|Z_{(k)} = j)$ using the corresponding sample variance, denoted as $\widehat{\sigma}_{kj}^2$, where $k = 1, \dots, K$ and $j = 1, 0$. Then we have an estimator for the standard error of $\widehat{\theta}(\mathcal{O})$,

$$\widehat{\text{SE}} = \left\{\sum_{k=1}^{K} \pi_k^2 \left[\frac{\widehat{\sigma}_{k1}^2}{N(k)g_{k1}} + \frac{\widehat{\sigma}_{k0}^2}{N(k)(1 - g_{k1})}\right]\right\}^{1/2}.$$

Thus, we can obtain both 95% CI for $\theta(P_{\mathcal{O}})$ and p-value for $H_0 : \theta(P_{\mathcal{O}}) = 0$ vs. $H_a : \theta(P_{\mathcal{O}}) \neq 0$.

2.5 Exercises

Ex 2.1

Download ICH E9 from the website, www.ich.org. Read Section 2; in particular, Subsection 2.3 "Design technique to avoid bias."

Ex 2.2

Consider a complete RCT. Following Fisher's approach, we treat potential outcomes, $(Y_i^{z=1}, Y_i^{z=0}), i = 1, \ldots, N$, as fixed values. Based on data (Z_i, Y_i) where $Y_i = Y_i^{z=Z_i}$, $i = 1, \ldots, N$, to test the sharp null hypothesis,

$$H_0 : Y_i^{z=1} = Y_i^{z=0}, \quad i = 1, \ldots, N,$$

consider the following test statistic,

$$T(Z_1, \ldots, Z_N) = \frac{\sum_{i=1}^N Z_i Y_i}{\sum_{i=1}^N Z_i} - \frac{\sum_{i=1}^N (1 - Z_i) Y_i}{\sum_{i=1}^N (1 - Z_i)}.$$

Design a plan of using the idea of permutation to obtain the p-value associated with the above the test statistic.

Ex 2.3

In R, generate a data frame that has 3 columns (subject ID `SID`, treatment assignment `Z` being 0 or 1, and observed outcome `Yobs`) and 100 rows, using the following R codes:

```
set.seed(3)
SID <- 1:100
Y0 <- rnorm(n = 100, mean = 0, sd = 1)
Y1 <- rnorm(n = 100, mean = 1, sd = 1.5)
Z <- rbinom(n = 100, size = 1, prob = 0.5)
Yobs <- Z*Y1 + (1-Z)*Y0
dataset3 <- data.frame(SID = SID, Z = Z, Yobs = Yobs)
```

Using simulation to generate a super population of $N^* = 10^6$ i.i.d. copies of $(Y^{z=1}, Y^{z=0})$, and have an approximate value of the causal estimand associated with the super population, denoted as θ^*.

Based on the above sample of size $N = 100$ that exists in the real world, obtain a 95% confidence interval estimate of θ^* using Neyman's approach.

Ex 2.4

In R, generate a data frame that has 3 columns (subject ID `SID`, treatment assignment `Z` being 0 or 1, and observed outcome `Yobs`) and 100 rows, using the following R codes:

```
set.seed(4)
SID <- 1:100
Y0 <- rbinom(n = 100, size = 1, prob = 0.3)
Y1 <- rbinom(n = 100, size = 1, prob = 0.7)
Z <- rbinom(n = 100, size = 1, prob = 0.5)
Yobs <- Z*Y1 + (1-Z)*Y0
dataset3 <- data.frame(SID = SID, Z = Z, Yobs = Yobs)
```

Using simulation to generate a super population of $N^* = 10^6$ i.i.d. copies of $(Y^{z=1}, Y^{z=0})$, and have an approximate value of the causal estimand associated with the super population, denoted as θ^*.

Based on the above sample of size $N = 100$ that exists in the real world, obtain a 95% confidence interval estimate of θ^* using Neyman's approach.

3

Missing Data Handling

3.1 Missing Data

"There is nothing known as 'Perfect'. Its only those imperfections which we choose not to see!!" —Albert Einstein

In Chapter 2, we discussed how to estimate the intent-to-treat (ITT) effect using randomized clinical trials (RCTs), which are assumed to be perfectly conducted. However, in reality, very likely we have to deal with imperfect RCTs in which some of the outcome measurements are missing. Therefore, we turn to a very impactful report titled "The Prevention and Treatment of Missing Data in Clinical Trials" (National Research Council 2010).

"Unfortunately, this key advantage, derived from the use of random selection for treatment and control groups, is jeopardized when some of the outcome measurements are missing."—National Research Council (2010)

After National Research Council (2010) pointed out the problem of missing data, they went on to provide the definition of missing data: "By missing data we mean when an outcome value that is meaningful for analysis is not collected."

How to handle missing data is a complex and difficult task. In this chapter, we focus on dropouts, a major source of missing data in clinical trials when participants discontinue their assigned treatments. In the next chapter, we consider other sources of missing data and other types of intercurrent events (ICEs), noting that dropouts are one type of ICEs.

National Research Council (2010) distinguished between **treatment dropouts** and **analysis dropouts**: "Treatment dropout is the result of a participant in a clinical trial discontinuing treatment; analysis dropout is the result of the failure to measure the outcome of interest for a trial participant."

Treatment dropout may lead to analysis dropout because the investigator may see no need to collect outcome data after a participant deviates from the study protocol. Moreover, terminal events such as death are not considered dropouts. For continuous or binary outcomes, death will be considered another type of ICE instead of dropout. For time-to-event outcome, if time-to-death

DOI: 10.1201/9781003433378-3

is the primary outcome, then death is considered the primary event in the definition of the time-to-event outcome; otherwise, death is considered as a competing event.

To prevent missing data due to treatment dropout, National Research Council (2010) provided 18 recommendations. For example, the 3rd recommendation is shown below.

> "Trial sponsors should continue to collect information on key outcomes on participants who discontinue their protocol-specified intervention in the course of the study, except in those cases for which a compelling cost-benefit analysis argues otherwise, and this information should be recorded and used in the analysis."—The 3rd recommendation in National Research Council (2010)

3.2 Intent-to-treat Effect

We consider two scenarios. Scenario one: the outcome data are still collected after treatment dropouts (i.e., there are treatment dropouts but no analysis dropouts). Scenario two: some outcome data are not collected after treatment dropouts (i.e., there are both treatment dropouts and analysis dropouts).

3.2.1 Scenario one

In this scenario, the outcome data are still collected even after the occurrence of a treatment dropout (e.g., discontinue the assigned treatment and receive no other treatment; discontinue the assigned treatment and then receive the standard of care or rescue medication).

Consider a population of size N to be enrolled in an RCT with a certain type of randomization, say block randomization. In the potential world, the population $\mathcal{P}$ would have potential outcomes $\{(Y_i^{z=1}, Y_i^{z=0}), i = 1, \ldots, N\}$. Furthermore, construct a super-population such that these N pairs of potential outcomes, $(Y_i^{z=1}, Y_i^{z=0})$'s, are i.i.d. with $(Y^{z=1}, Y^{z=0})$. Because these potential outcomes exist in the potential world before the study is conducted, they are the same either for the perfect RCT without any dropout or for any imperfect RCT with treatment dropouts.

As in Chapter 2, if we are interested in the ITT effect, the causal estimand of interest associated with the super-population is

$$\theta^*_{\text{ITT}} = \mathbb{E}\left(Y^{z=1} - Y^{z=0}\right) \tag{3.1}$$

for continuous or binary outcome variables.

In this scenario, data after treatment dropouts are still collected, so we have observed data $\mathcal{O} = \{(Z_i, Y_i), i = 1, \ldots, N\}$, where (Z_i, Y_i)'s are identically distributed with (Z, Y). Following the same arguments in Chapter 2, we can translate causal estimand (3.1) into a statistical estimand,

$$\theta_{\text{ITT}} = \mathbb{E}(Y|Z = 1) - \mathbb{E}(Y|Z = 0), \tag{3.2}$$

under the consistency assumption that $Y_i = Y_i^{z=Z_i}, i = 1, \ldots, N$.

Following the same arguments in Chapter 2, we can show that

$$\widehat{\theta}_{\text{ITT}} = \frac{\sum_{i=1}^N Z_i Y_i}{\sum_{i=1}^N Z_i} - \frac{\sum_{i=1}^N (1 - Z_i) Y_i}{\sum_{i=1}^N (1 - Z_i)} \tag{3.3}$$

is an unbiased estimator of θ_{ITT}. Furthermore, $\widehat{\theta}_{\text{ITT}}$ is an unbiased estimator of θ^*_{ITT} under the consistency assumption.

3.2.2 Scenario two

In this scenario, not all the outcome data after treatment dropouts are collected—some treatment dropouts lead to analysis dropouts, along with other analysis dropouts that are not due to treatment dropouts (e.g., withdraw from the study for reasons not related to treatment).

In the potential world, the population $\mathcal{P}$ would consist of potential outcomes $\{(Y_i^{z=1}, Y_i^{z=0}), i = 1, \ldots, N\}$, which are i.i.d. with $(Y^{z=1}, Y^{z=0})$. Because these potential outcomes exist in the potential world before the study is conducted, they are the same either for the perfect RCT without dropouts or any imperfect RCT with dropouts. Similar to scenario one, in scenario two we may be also interested in the ITT estimand defined in (3.1).

After an RCT is conducted, we obtain data $\mathcal{O} = \{(X_i, Z_i, \Delta_i, (1 - \Delta_i)Y_i), i = 1, \ldots, N\}$, where X_i is a vector of baseline variables, Z_i is treatment assignment, $\Delta_i = 1$ if outcome Y_i is missing due to analysis dropout and $\Delta_i = 0$ if Y_i is observed, and $(1 - \Delta_i)Y_i$ is the observed outcome. Assume that $(X_i, Z_i, \Delta_i, Y_i)$'s are identically distributed with (X, Z, Δ, Y). See Table 3.1 for an example of such dataset.

Rubin (1976) gave a taxonomy for missing data: missing completely at random (MCAR), missing at random (MAR), and missing not at random (MNAR). In the case of MCAR, the missing data are unrelated to the study variables. In the case of MAR, whether or not data are missing may depend on the values of the observed study variables. In the case of MNAR, whether or not data are missing may depend on the values of the missing data.

A common practice is to perform analysis under the MAR assumption along with other identifiability assumptions, and then conduct sensitivity analysis to explore the robustness of the main results under the MNAR assumption. For the main analysis, we make the following three assumptions:

TABLE 3.1
Data for estimating the ITT effect

SID	X	Z	Δ	$(1-\Delta)Y$
$\vdots$	$\vdots$	$\vdots$	$\vdots$	$\vdots$
i_1	X_{i_1}	1	0	Y_{i_1}
i_2	X_{i_2}	1	1	NA*
i_3	X_{i_3}	0	0	Y_{i_3}
i_4	X_{i_4}	0	1	NA
$\vdots$	$\vdots$	$\vdots$	$\vdots$	$\vdots$

* NA stands for missing value

1. The consistency assumption:

$$Y = Y^{z=j} \text{ if } Z = j \text{ and } \Delta = 0, \text{ for } j = 1, 0; \tag{3.4}$$

2. The MAR assumption:

$$(Y^{z=0}, Y^{z=1}) \perp\!\!\!\perp \Delta | (X, Z); \tag{3.5}$$

3. The positivity assumption:

$$\mathbb{P}(\Delta = 0 | Z = j, X = x) > 0, \text{ for } j = 1, 0; x \in \text{supp}(X). \tag{3.6}$$

Remark 1: In a study where there is only one follow-up visit, the MAR assumption (3.5) is also called the covariate-dependent MCAR assumption (O'Kelly and Ratitch 2014). In a longitudinal study where there are multiple follow-up visits, the MAR assumption will become more relaxing than covariate-dependent MCAR.

Remark 2: To visualize the MAR assumption (3.5), we consider the directed acyclic graph (DAG) in Figure 3.1 to display the relations among $\{X, Z, \Delta, Y\}$. In the DAG, path $X \to Z$ is added to cover the case when the stratification factor S is part of X, and Δ depends on X and Z. As demonstrated in Appendix 3.5 using the tool of the structural causal model (SCM) (Pearl 2009), this DAG implies that $(Y^{z=1}, Y^{z=0}) \perp\!\!\!\perp \Delta | (X, Z)$.

Remark 3: In the positivity assumption (3.6), $x \in \text{supp}(X)$ stands for that x is in the support of X; that is, the mass probability function or density of X, denoted as $p_X(x)$, is positive at x.

In order to propose an estimator to estimate (3.1), we need to translate the causal estimand into a statistical estimand under the above three assumptions.

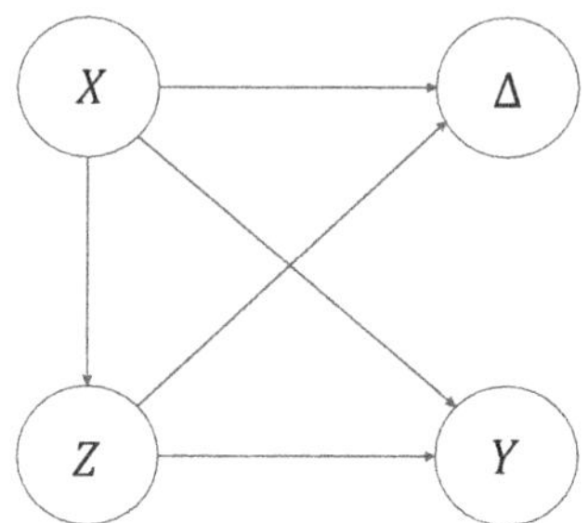

FIGURE 3.1
DAG that illustrates MAR (or covariate-dependent MCAR)

For this aim, we consider the following proof:

$$
\begin{aligned}
\theta^*_{\text{ITT}} &= \mathbb{E}(Y^{z=1}) - \mathbb{E}(Y^{z=0}) \\
&\overset{(a)}{=} \mathbb{E}[\mathbb{E}(Y^{z=1}|X)] - \mathbb{E}[\mathbb{E}(Y^{z=0}|X)] \\
&\overset{(b)}{=} \mathbb{E}[\mathbb{E}(Y^{z=1}|Z=1,X)] - \mathbb{E}[\mathbb{E}(Y^{z=0}|Z=0,X)] \\
&\overset{(c)}{=} \mathbb{E}[\mathbb{E}(Y^{z=1}|\Delta=0,Z=1,X) - \mathbb{E}(Y^{z=0}|\Delta=0,Z=0,X)] \\
&\overset{(d)}{=} \mathbb{E}[\mathbb{E}(Y|\Delta=0,Z=1,X) - \mathbb{E}(Y|\Delta=0,Z=0,X)] \\
&\overset{(e)}{=} \mathbb{E}[Q_{\Delta=0}(Y|1,X) - Q_{\Delta=0}(Y|0,X)] \triangleq \theta_{\text{ITT}},
\end{aligned}
$$

where (a) holds using the law of iterated expectation, (b) holds because of the randomization (either complete randomization or stratified randomization provided that X includes the stratification factor S), (c) holds under the MAR assumption and the positivity assumption, (d) holds under the consistency assumption, and in (e),

$$Q_{\Delta=0}(Z,X) = \mathbb{E}(Y|\Delta=0,Z,X) \tag{3.7}$$

is the regression function of Y against (Z,X) given $\Delta = 0$.

Based on the above proof, the resulting θ_{ITT} is the statistical estimand that is equal to the causal estimand under the consistency, MAR, and positivity assumptions—these assumptions are referred to as the identifiability assumptions (Hernán and Robins 2020).

Hence, if we can obtain an estimator for $Q_{\Delta=0}(Z,X)$, denoted as $\widehat{Q}_{\Delta=0}(Z,X)$, based on the observed data through some regression method (say, linear regression for continuous outcome, logistic regression for binary outcome, or any predictive modeling method), then we can construct an estimator for θ_{ITT},

$$\widehat{\theta}_{\text{ITT}} = \frac{1}{N}\sum_{i=1}^{N}\left[\widehat{Q}_{\Delta=0}(1,X_i) - \widehat{Q}_{\Delta=0}(0,X_i)\right]. \tag{3.8}$$

We relegate the discussion on whether there are better estimation methods for estimating the statistical estimand and how to estimate their variances to Chapters 7–9. In this chapter and then Chapters 4–6, we focus on the task of identification from causal estimand to statistical estimand, along with a simple demonstration that there is at least one method—regardless of how good it is—to estimate the statistical estimand.

We conclude this section with some discussion on the *complete-case analysis*, which is based on the subset $\{i : \Delta_i = 0, i = 1, \ldots, N\}$. The complete-case analysis is valid under the MCAR assumption (i.e., missing status is independent of covariates and treatment group), which is stricter than the MAR assumption. Since the methods under the MAR assumption are well-developed, including the method of (3.8) that is based on all the observed data, we do not recommend the complete-case analysis.

3.3 Per-protocol Effect

Consider a population of size N to be enrolled in an RCT. In the potential world, we could imagine as many potential outcomes as possible; for example, we could imagine a population $\mathcal{P}$ consisting of potential outcomes $\{(Y_i^{z=1}, Y_i^{z=0}, Y_i^{z=1,a=1}, Y_i^{z=1,a=0}, Y_i^{z=1,a=\texttt{NULL}}, Y_i^{z=1,a=2}, Y_i^{z=0,a=0}, Y_i^{z=0,a=1}, Y^{z=0,a=\texttt{NULL}}, Y^{z=0,a=2}), i = 1, \ldots, N\}$.

Here $Y_i^{z=j_1,a=j_2}$ is the potential outcome had the subject i been assigned to treatment j_1, which could be either treatment 1 or treatment 0, and then been treated by treatment j_2, which could be treatment 1, treatment 0, no treatment (denoted as `NULL`), or another treatment (referred to as treatment 2). In particular, $Y_i^{z=j,a=j}$ is the potential outcome had the subject i been assigned to treatment j and then treated by treatment j per protocol.

Figure 3.2 displays the DAG that illustrates the relationships among assignment variable Z, treatment variable A, analysis dropout status Δ, and outcome variable Y. If the RCT is double-blind, then we can remove the arrow from Z to Y, implying $Y^{z=j_1,a=j_2} = Y^{a=j_2}$, because the blinding ensures that the effect of Z on Y is only through A. On the other hand, if the RCT is not double-blind, then $Y^{z=j_1,a=j_2}$ might be different from $Y^{a=j_2}$.

Here are definitions of D, Δ, and $\widetilde{\Delta}$ shown in Figure 3.2. Let $D = I(Z \neq A)$ be the indicator of protocol deviation for a subject, with $D = 0$ standing for no protocol deviation (i.e., per protocol). As in Section 3.2, let Δ be the indicator of analysis dropout. Since outcome data may be still collected even if there is protocol deviation, $D = 1$ does not imply $\Delta = 1$. Moreover, $D = 0$ and $\Delta = 0$ means there is no protocol deviation and the outcome is observed, while $D = 0$ and $\Delta = 1$ means there is no protocol deviation but the outcome is missing. Finally, let $\widetilde{\Delta}$ be the indicator of dropout (either treatment dropout

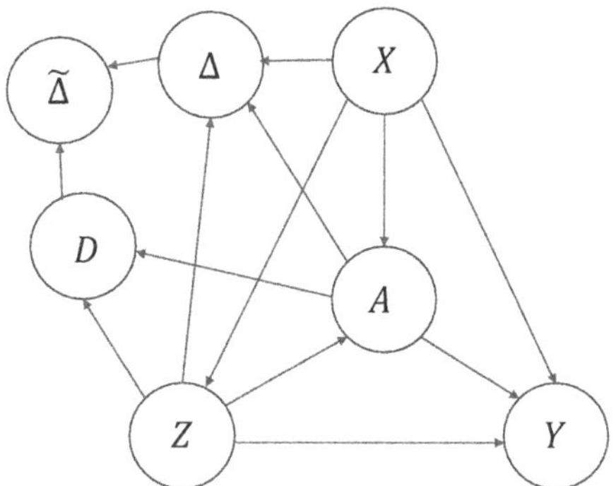

FIGURE 3.2
DAG that illustrates treatment dropouts and analysis dropouts

or analysis dropout); that is,

$$\widetilde{\Delta} = 1, \text{ if } D = 1 \text{ or } \Delta = 1;$$
$$\widetilde{\Delta} = 0, \text{ if } D = 0 \text{ and } \Delta = 0.$$

By these definitions, D is a function of Z and A—hence there are arrows $Z \to D$ and $A \to D$ in the DAG; and $\widetilde{\Delta}$ is a function of D and Δ—hence there are arrows $D \to \widetilde{\Delta}$ and $\Delta \to \widetilde{\Delta}$ in the DAG.

Therefore, if we are interested in the per-protocol (PP) effect, the estimand of interest is defined as

$$\theta^*_{\text{PP}} = \mathbb{E}\left(Y^{z=1,a=1} - Y^{z=0,a=0}\right) \tag{3.9}$$

for continuous or binary outcome variables.

The amount of missing data or "missing" data—with the quotation marks emphasizing the broader definition of missing data—depends on whether the ITT estimand or the PP estimand is of interest. If θ^*_{ITT} is of interest, the missing rate is

$$r = \frac{1}{N}\sum_{i=1}^{N} \Delta_i, \tag{3.10}$$

because the data that are collected after treatment dropouts are considered relevant and only the analysis dropouts lead to missing data. However, in the estimation of the PP estimand, the data after treatment dropouts—regardless of being collected or not—are considered irrelevant. Therefore, if θ^*_{PP} is of interest, the "missing" rate is

$$\widetilde{r} = \frac{1}{N}\sum_{i=1}^{N} \widetilde{\Delta}_i, \tag{3.11}$$

TABLE 3.2
Data for estimating the PP effect

SID	X	Z	A	D	Δ	$\widetilde{\Delta}$	$(1-\widetilde{\Delta})Y$
$\vdots$	$\vdots$	$\vdots$	$\vdots$	$\vdots$	$\vdots$	$\vdots$	$\vdots$
i_1	X_{i_1}	1	1	0	0	0	Y_{i_1}
i_2	X_{i_2}	1	1	0	1	1	NA
i_3	X_{i_3}	1	$\neq 1$	1	0	1	NA
i_4	X_{i_4}	1	$\neq 1$	1	1	1	NA
i_5	X_{i_5}	0	0	0	0	0	Y_{i_5}
i_6	X_{i_6}	0	0	0	1	1	NA
i_7	X_{i_7}	0	$\neq 0$	1	0	1	NA
i_8	X_{i_8}	0	$\neq 0$	1	1	1	NA
$\vdots$	$\vdots$	$\vdots$	$\vdots$	$\vdots$	$\vdots$	$\vdots$	$\vdots$

Thus, to estimate the PP estimand, we will use data $\mathcal{O} = \{(X_i, Z_i, A_i, D_i, \Delta_i, \widetilde{\Delta}_i, (1-\widetilde{\Delta}_i)Y_i), i = 1, \ldots, N\}$. See Table 3.2 for an example of such a dataset.

A common practice is to perform analysis under the "missing" at random (also abbreviated as MAR, to abuse the abbreviation) assumption along with other identifiability assumptions, and then conduct sensitivity analysis to explore the robustness of the main results under the "missing" not at random (also abbreviated as MNAR, to abuse the abbreviation) assumption. Thus, for the main analysis, we make the following three assumptions:

1. The consistency assumption:

$$Y = Y^{z=j,a=j} \text{ if } Z = j \text{ and } \widetilde{\Delta} = 0, \text{ for } j = 1, 0; \tag{3.12}$$

2. The MAR assumption:

$$(Y^{z=1,a=1}, Y^{z=0,a=0}) \perp\!\!\!\perp \widetilde{\Delta} | (X, Z); \tag{3.13}$$

3. The positivity assumption:

$$\mathbb{P}(\widetilde{\Delta} = 0 | Z = j, X = x) > 0, \text{ for } j = 1, 0; x \in \text{supp}(X). \tag{3.14}$$

To visualize the MAR assumption (3.13), we consider the DAG in Figure 3.2. With the help of SCMs in Appendix 3.5, the DAG in Figure 3.2 implies that $(Y^{z=1,a=1}, Y^{z=0,a=0}) \perp\!\!\!\perp \widetilde{\Delta} | (X, Z)$.

In order to propose an estimator to estimate (3.9), we need to translate the causal estimand into a statistical estimand under the above three assumptions. Following the same arguments in Section 3.2, except that $\Delta = 0$ is replaced by $\widetilde{\Delta} = 0$, we can show that

$$\begin{aligned}\theta^*_{\mathrm{PP}} =& \mathbb{E}\left(Y^{z=1,a=1} - Y^{z=0,a=0}\right)\\ =& \mathbb{E}[\mathbb{E}(Y|\widetilde{\Delta} = 0, Z = 1, X) - \mathbb{E}(Y|\widetilde{\Delta} = 0, Z = 0, X)]\\ =& \mathbb{E}[Q_{\widetilde{\Delta}=0}(Y|1, X) - Q_{\widetilde{\Delta}=0}(Y|0, X)] \triangleq \theta_{\mathrm{PP}},\end{aligned}$$

where

$$Q_{\widetilde{\Delta}=0}(Z, X) = \mathbb{E}(Y|\widetilde{\Delta} = 0, Z, X), \tag{3.15}$$

is the regression function of Y against (Z, X) given $\widetilde{\Delta} = 0$.

If we can have an estimator for $Q_{\widetilde{\Delta}=0}(Z, X)$, denoted as $\widehat{Q}_{\widetilde{\Delta}=0}(Z, X)$, based on the observed data through some method (say, generalized linear model), then we can have an estimator for θ_{PP},

$$\widehat{\theta}_{\mathrm{PP}} = \frac{1}{N}\sum_{i=1}^{N}\left[\widehat{Q}_{\widetilde{\Delta}=0}(1, X_i) - \widehat{Q}_{\widetilde{\Delta}=0}(0, X_i)\right]. \tag{3.16}$$

We conclude this section with some discussion on the *naive per-protocol analysis*, which is based on the subset with $\{i : \widetilde{\Delta}_i = 0, i = 1, \ldots, N\}$. The naive per-protocol analysis is valid under the "missing" completely at random assumption (i.e., "missing" status is independent of covariates and treatment group), which is stricter than the MAR assumption. Therefore, we do not recommend the naive per-protocol analysis.

3.4 Sources of Missing Data

3.4.1 Intercurrent events

In the above two sections, we focus on dropouts and distinguish between treatment dropouts and analysis dropouts. The amount of missing data or "missing" data depends on how we handle dropouts and whether we are interested in the ITT effect or the PP effect. Since dropouts are events occurring after treatment initiation that affect the existence of the measurements associated with the estimand—either ITT or PP, dropouts are intercurrent events (ICEs), according to the following definition in ICH E9(R1):

> "Events occurring after treatment initiation that affect either the interpretation or the existence of the measurements associated with the clinical question of interest. It is necessary to address intercurrent events when describing the clinical question of interest in order to precisely define the treatment effect that is to be estimated."—ICH E9(R1)

In Section 3.2, we are interested in the ITT estimand. As indicated by Δ, the missing data are consequences of analysis dropouts. In the definition of the ITT estimand in (3.1), we apply the so-called **treatment policy strategy**, one of the five strategies in ICH E9(R1) to handle treatment dropouts.

In Section 3.3, we are interested in the PP estimand. Indicated by $\widetilde{\Delta}$, the "missing" data are consequences of treatment dropouts and/or analysis dropouts. In the definition of the PP estimand in (3.9), we apply the so-called **hypothetical strategy**, one of the five strategies in ICH E9(R1) to handle treatment dropouts. Note that, by the hypothetical strategy, the data collected after treatment dropouts are considered as "missing" data, with quotation marks emphasizing that those data are existing in the real world but considered as missing in the hypothetical world because they are not relevant to the PP effect.

3.4.2 Missing data that are consequences of ICEs

In the protocol development, we need to specify the strategies for handling all the anticipated ICEs and the strategies for handling all the anticipated missing data. Therefore, we need to distinguish between (i) missing data that are consequences of ICEs and (ii) missing data that are not consequences of ICEs. For missing data that are consequences of ICEs, they can be handled automatically using the strategies that handle their ICE sources. For missing data that are not consequences of ICEs, separate strategies for handling them should be specified.

In this subsection, we discuss how to handle missing data that are consequences of ICEs. First, we need to categorize all the anticipated ICEs that may lead to missing data into several categories; for example,

- Treatment discontinue due to early efficacy,
- Treatment discontinue due to lack of efficacy,
- Treatment discontinue due to non-health related reasons,
- Treatment add-on,
- Switching to a different treatment due to heath related reasons,
- Switching to a different treatment due to non-health related reasons,
- Analysis dropout due to health related reasons,
- Analysis dropout due to non-health related reasons,
- Terminal event such as death,
- Administrative censoring for time-to-event outcome.

Next, we need to specify a strategy or a combination of several strategies for handling the categorized ICEs. These ICH handling strategies will be discussed in more detail in the next chapter. How to handle ICEs is one of the five attributes of the estimand of interest. Consequently, study planning, design, and conduct should be aligned with the estimand.

Finally, we need to clarify whether or not all the anticipated missing data can be handled by the above strategies. If not, then there are missing data that are not consequences of ICEs; see the next subsection for a discussion on how to handle them.

3.4.3 Missing data that are not consequences of ICEs

For missing data that are not consequences of certain ICEs, it is unnecessary to address them when describing the clinical question of interest and defining the estimand of interest, although we still need to describe the methods to handle them in the protocol.

Here are some examples of such missing data. Example 1: There are missing data in covariates. Example 2: In outcomes that are composites of multiple variables, some of which have missing data. Example 3: In longitudinal studies, some intermediate visits are missing while data on the final visits are available.

How to handle such missing data falls out of the main focus of this book, while we will briefly discuss how to handle them in the following chapters whenever it is ready. For example, in Chapter 8, we will discuss how to handle missing data in baseline covariates.

We conclude this section with ICH E9(R1)'s definition of missing data:

> "Data that would be meaningful for the analysis of a given estimand but were not collected. They should be distinguished from data that do not exist or data that are not considered meaningful because of an intercurrent event."—The definition of missing data in ICH E9(R1).

In the above definition, three types of data are mentioned:

- Data that would be meaningful for the analysis of a given estimand but were not collected—these data are considered missing data.

- Data that are not considered meaningful because of an intercurrent event. For example, these data that are still collected after treatment dropouts are considered as "missing" data if the hypothetical strategy is applied to handle treatment dropouts, where the quotation marks are used to distinguish this type of data with the missing data.

- Data that do not exist. For example, since mammogram screening is only performed on female participants, the data on mammogram screening do not exist for male participants. Thus, those data that do not exist are neither missing data nor "missing" data.

3.5 Appendix

A DAG is equivalent to a series of structural causal models (SCMs). Variables in the DAG are called endogenous variables, while variables that are not in the DAG are called exogenous variables. Exogenous variables are assumed to be independent. Each endogenous variable corresponds to an unknown deterministic function of its parents and one exogenous variable (Pearl 2009).

Take the DAG in Figure 3.1 as an example. We have the following structural causal models for endogenous variables X, Z, Δ, and Y:

$$\begin{aligned}X &= f_X(U_X),\\ Z &= f_Z(X, U_Z),\\ \Delta &= f_\Delta(X, Z, U_\Delta),\\ Y &= f_Y(X, Z, U_Y),\end{aligned}$$

where $U.$'s are exogenous variables, which are independent of one other, and $f.$'s are deterministic functions. Using SCMs, we can express potential outcomes $Y^{z=1}$ and $Y^{z=0}$ explicitly:

$$\begin{aligned}Y^{z=1} &= f_Y(X, 1, U_Y),\\ Y^{z=0} &= f_Y(X, 0, U_Y).\end{aligned}$$

Because U_Δ and U_Y are independent, $(f_Y(X,1,U_Y), f_Y(X,0,U_Y))$ and $f_\Delta(X,Z,U_\Delta)$ are independent given (X,Z). That is, $(Y^{z=1}, Y^{z=0}) \perp\!\!\!\perp \Delta | (X, Z)$.

Furthermore, $f_\Delta(X,Z,U_\Delta) = f_\Delta(X, f_Z(X,U_Z), U_\Delta)$. Because U_Y and (U_Z, U_Δ) are independent, we see that $(f_Y(X,1,U_Y), f_Y(X,0,U_Y))$ and $f_\Delta(X, f_Z(X,U_Z), U_\Delta)$ are independent given X. That is, $(Y^{z=1}, Y^{z=0}) \perp\!\!\!\perp \Delta | X$.

Take the DAG in Figure 3.2 as another example. We have the following SCMs for endogenous variables X, Z, A, Δ, D, $\widetilde{\Delta}$, and Y:

$$\begin{aligned}X &= f_X(U_X),\\ Z &= f_Z(X, U_Z),\\ A &= f_A(X, Z, U_A),\\ \Delta &= f_\Delta(X, Z, A, U_\Delta),\\ D &= f_D(Z, A),\\ \widetilde{\Delta} &= f_{\widetilde{\Delta}}(\Delta, D),\\ Y &= f_Y(X, Z, A, U_Y),\end{aligned}$$

where $U.$'s are exogenous variables, which are independent of each other, and $f.$'s are deterministic functions. Using SCMs, we can express potential outcomes $Y^{z=1,a=1}$ and $Y^{z=0,a=1}$ explicitly:

$$\begin{aligned}Y^{z=1,a=1} &= f_Y(X, 1, 1, U_Y),\\ Y^{z=0,a=0} &= f_Y(X, 0, 0, U_Y).\end{aligned}$$

Because U_Y and (U_Δ, U_A) are independent, $(f_Y(X,1,1,U_Y), f_Y(X,0,0,U_Y))$ and $f_\Delta(X,Z,A,U_\Delta) = f_\Delta(X,Z,f_A(X,Z,U_A),U_\Delta)$ are independent given (X,Z). That is, $(Y^{z=1,a=1}, Y^{z=0,a=0}) \perp\!\!\!\perp \Delta|(X,Z)$. In addition, because U_Y and U_A are independent, $(f_Y(X,1,1,U_Y), f_Y(X,0,0,U_Y))$ and $f_D(Z,A) = f_D(Z, f_A(X,Z,U_A)$ are independent given (X,Z). That is, $(Y^{z=1,a=1}, Y^{z=0,a=0}) \perp\!\!\!\perp D|(X,Z)$. Combining these two results, $(Y^{z=1,a=1}, Y^{z=0,a=0}) \perp\!\!\!\perp \widetilde{\Delta}|(X,Z)$.

Furthermore, if we replace Z by $f_Z(X,U_Z)$ in the above statements that involve Z, we can show that $(Y^{z=1,a=1}, Y^{z=0,a=0}) \perp\!\!\!\perp \widetilde{\Delta}|X$ as well.

3.6 Exercises

Ex 3.1

Read those 18 recommendations provided in *The Prevention and Treatment of Missing Data in Clinical Trials* (National Research Council 2010) available at https://www.ncbi.nlm.nih.gov/books/NBK209904/.

Ex 3.2

In Section 3.2, we use the standardization strategy to translate the causal estimand θ^*_{ITT} into a statistical estimand θ_{ITT}. Alternately, we can use the weighting strategy to do the job. That is, we can show that

$$
\begin{aligned}
\theta^*_{\text{ITT}} =& \mathbb{E}(Y^{z=1}) - \mathbb{E}(Y^{z=0}) \\
\overset{(a)}{=}& \mathbb{E}\{\mathbb{E}(Y^{z=1}|X)\} - \mathbb{E}\{\mathbb{E}(Y^{z=0}|X)\} \\
\overset{(b)}{=}& \mathbb{E}\{\mathbb{E}(Y^{z=1}|Z=1,X)\} - \mathbb{E}\{\mathbb{E}(Y^{z=0}|Z=0,X)\} \\
\overset{(c)}{=}& \mathbb{E}\left\{\mathbb{E}\left[\left.\frac{I(\Delta=0)Y^{z=1}}{\mathbb{P}(\Delta=0|Z=1,X)}\right| Z=1,X\right]\right\} \\
& - \mathbb{E}\left\{\mathbb{E}\left[\left.\frac{I(\Delta=0)Y^{z=0}}{\mathbb{P}(\Delta=0|Z=0,X)}\right| Z=0,X\right]\right\} \\
\overset{(d)}{=}& \mathbb{E}\left\{\mathbb{E}\left[\left.\frac{I(\Delta=0)Y}{\mathbb{P}(\Delta=0|Z=1,X)}\right| Z=1,X\right]\right\} \\
& - \mathbb{E}\left\{\mathbb{E}\left[\left.\frac{I(\Delta=0)Y}{\mathbb{P}(\Delta=0|Z=0,X)}\right| Z=0,X\right]\right\} \\
\triangleq& \theta_{\text{ITT}}.
\end{aligned}
$$

Write down the reasons why equations (a)–(d) hold.

Ex 3.3

In R, generate a data frame that has 5 columns (subject ID `SID`, covariate `X`, treatment assignment `Z` being 0 or 1, missing indicator `Delta`, and observed incomplete outcome `Yobs`) and 100 rows, using the following R codes:

```
set.seed(5)
SID <- 1:100
Y0 <- rnorm(n=100, mean=0, sd=1)
Y1 <- rnorm(n=100, mean=1, sd=1.5)
X <- rnorm(n=100, mean=0, sd=2)
Z <- rbinom(n=100 , size=1, prob=0.5)
Delta <- rbinom(n=100, size=1, prob=exp(-2+X)/(1+exp(-2+X)))
mean(Delta) # missing rate
Yobs <- ifelse(Delta, NA, Z*Y1+(1-Z)*Y0)
summary(Yobs) # summary of variable with missing data
dataset5 <- data.frame(SID=SID, X=X, Z=Z, Delta=Delta, Yobs=Yobs)
```

Using simulation to generate a super-population of $N^* = 10^6$ i.i.d. copies of $(Y^{z=1}, Y^{z=0})$, obtain an approximate value of the causal ITT estimand, denoted as θ^*_{ITT}. Based on the above dataset of $N = 100$ with missing data, obtain a point estimate of θ^*_{ITT} using the method discussed in Section 3.2.

4

Intercurrent Events Handling

4.1 Five Strategies

In Chapter 2, we discussed how to estimate the intent-to-treat (ITT) effect using randomized clinical trials (RCTs) that are perfectly conducted—no missing data and no intercurrent events (ICEs). In a perfect RCT, the ITT effect is the same as the per-protocol (PP) effect. In reality, most likely we encounter imperfect RCTs with missing data or ICEs. In Chapter 3, we discussed how to handle missing data that are consequences of ICEs. In this chapter, we return to the sources of missing data—ICEs—and discuss how to handle them.

As recommended in ICH E9(R1), "it is necessary to address intercurrent events when describing the clinical question of interest in order to precisely define the treatment effect that is to be estimated."

On the one hand, in Chapter 1, we discussed the PROTECT checklist for describing the clinical question of interest: Population, Response/Outcome, Treatment/Exposure, Counterfactual thinking, and Time. On the other hand, ICH E9(R1) summarized five strategies for addressing ICEs when defining the clinical question of interest: the treatment policy strategy, the hypothetical strategy, the composite variable strategy, the while on treatment strategy, and the principal stratum strategy.

There is a one-to-one correspondence between the five components of the PROTECT checklist and the five ICE handling strategies (Fang and He 2023a), as shown in Table 4.1.

In each of the following five subsections, we discuss one ICE handling strategy, assuming that there is only one category of ICEs to be handled by that strategy. After we discuss them individually, we will discuss how to apply several strategies to handle several types of ICEs simultaneously. Moreover, in this section we only consider continuous and binary outcomes, leaving the time-to-event outcomes to the next section.

DOI: 10.1201/9781003433378-4

4.1.1 The treatment policy strategy

Estimand

Assume that there is only one category of non-terminal ICEs (e.g., treatment dropouts such as treatment discontinuation and rescue medication) and the data after an ICE occurrence are still collected.

Let's read the description of the treatment policy strategy in ICH E9(R1):

> "The occurrence of the intercurrent event is considered irrelevant in defining the treatment effect of interest; the values for the variable of interest are used regardless of whether the intercurrent event occurs. For example, when specifying how to address the use of additional medication as an intercurrent event, the values of the variable of interest should be used, whether the patient takes additional medication or not."—ICH E9(R1)

To understand this strategy, we review the definitions of simple treatment, treatment regimen, and treatment policy that were defined in National Research Council (2010). *Simple treatment* would usually consist of a prescribed dose of a given frequency. *Treatment regimen* would usually involve rules for dose escalation or reduction in order to obtain greater effect while avoiding intolerable adverse experiences. *Treatment policy* would usually include plans for auxiliary treatments and progression to other treatments in the face of disease progression.

Let E be the indicator of ICE occurrence, where $E = 1$ means an ICE occurs and $E = 0$ means no ICE occurs. In the potential world, let $E^{z=j}$ be the potential ICE occurrence indicator had the patient been assigned to treatment j, $j = 1, 0$. Using the treatment policy strategy, the ICE E becomes a part of the treatment policy being compared, and therefore $Z = j$ means the patient is assigned to treatment j initially and then followed by $E^{z=j}$. For example, if the ICE under consideration is taking rescue medication due to lack of efficacy, then treatment policy $Z = j$ means that the patient initiates the treatment j, which may or may not be followed by rescue medication depending on whether the treatment is ineffective or effective.

Recall that we could imagine as many potential outcomes as possible in the potential world, as long as we put as many feasible interventions as possible to the superscript. Before the study is conducted, in the potential world, for each patient i, we imagine the following four potential outcomes:

$$Y_i^{z=1,e=0}, \text{ if patient } i \text{ were assigned to } Z = 1 \text{ and intervened by } E = 0,$$
$$Y_i^{z=1,e=1}, \text{ if patient } i \text{ were assigned to } Z = 1 \text{ and intervened by } E = 1,$$
$$Y_i^{z=0,e=0}, \text{ if patient } i \text{ were assigned to } Z = 0 \text{ and intervened by } E = 0,$$
$$Y_i^{z=0,e=1}, \text{ if patient } i \text{ were assigned to } Z = 0 \text{ and intervened by } E = 1.$$

TABLE 4.1
Correspondence between PROTECT and ICE handling strategies

Symbol	Checklist	ICE handling strategy
P	Population	Principal stratum strategy
R/O	Response/Outcome	Composite variable strategy
T/E	Treatment/Exposure	Treatment policy strategy
C	Counterfactual thinking	Hypothetical strategy
T	Time	While on treatment strategy

Thus, if there is only one type of ICE, we have

$$\begin{aligned} Y_i^{z=1} =& Y_i^{z=1,e=E^{z=1}}, \\ Y_i^{z=0} =& Y_i^{z=0,e=E^{z=0}}. \end{aligned}$$

Therefore, by the treatment policy strategy, comparing assignment $Z = 1$ versus assignment $Z = 0$ is equivalent to comparing treatment policy "$Z = 1, E = E^{z=1}$" versus treatment policy "$Z = 0, E = E^{z=0}$." That is, applying the treatment policy strategy is equivalent to revising the "T/E" component of the PROTECT checklist, as presented in Table 4.1.

To summarize, if there is only one type of ICE and we want to apply the treatment policy strategy to handle it, we are interested in the following causal estimand,

$$\theta_{\text{ITT}}^* = \mathbb{E}(Y^{z=1,e=E^{z=1}} - Y^{z=0,e=E^{z=0}}) = \mathbb{E}(Y^{z=1} - Y^{z=0}). \tag{4.1}$$

Estimator

Assuming that data after the ICE occurrence are still collected and therefore no missing data, we have data $\mathcal{O} = \{(Z_i, Y_i), i = 1, \ldots, N\}$. Following the same arguments in Chapter 3, we construct the following estimator,

$$\widehat{\theta}_{\text{ITT}} = \frac{\sum_{i=1}^N Z_i Y_i}{\sum_{i=1}^N Z_i} - \frac{\sum_{i=1}^N (1 - Z_i) Y_i}{\sum_{i=1}^N (1 - Z_i)}, \tag{4.2}$$

which is an unbiased estimator for θ_{ITT}^* under the consistency assumption.

Missing data handling

Assuming that there are missing data, we have data $\mathcal{O} = \{(X_i, Z_i, \Delta_i, (1 - \Delta_i)Y_i), i = 1, \ldots, N\}$, where X_i is a vector of baseline variables, Z_i is the treatment assignment, $\Delta_i = 1$ if Y_i is missing and $\Delta_i = 0$ if Y_i is observed, and $(1-\Delta_i)Y_i$ is the observed outcome. Following the same arguments in Chapter 3, under the identifiability assumptions (consistency, missing at

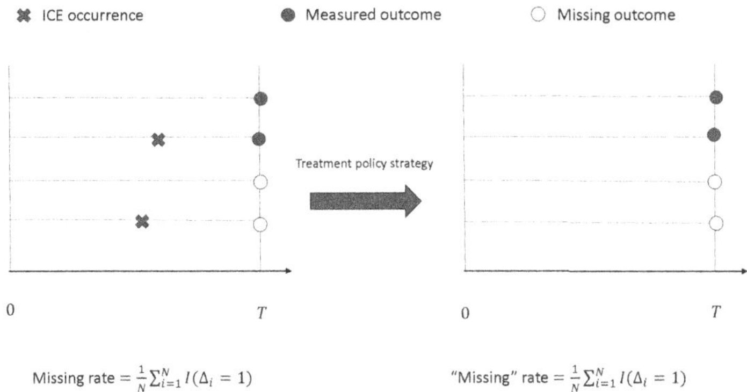

FIGURE 4.1
An illustration of the treatment policy strategy

random, positivity), we construct the following estimator,

$$\widehat{\theta}_{\text{ITT}} = \frac{1}{N}\sum_{i=1}^{N}\left[\widehat{Q}_{\Delta=0}(1, X_i) - \widehat{Q}_{\Delta=0}(0, X_i)\right], \tag{4.3}$$

where $\widehat{Q}_{\Delta=0}(Z, X)$ is an estimator of $Q_{\Delta=0}(Z, X) = \mathbb{E}(Y|\Delta = 0, Z, X)$.

Figure 4.1 illustrates the treatment policy strategy. There are four hypothetical subjects, one has observed outcome without any ICE, one has observed outcome after an ICE, one has missing outcome without ICE, and one has missing outcome after an ICE. By the treatment policy strategy, the values for the observed outcomes are used regardless of whether the ICE occurs. Therefore, the "missing" rate after the application of the treatment policy strategy is the same as the missing rate in the data.

4.1.2 The hypothetical strategy

Estimand

Assume that there is only one category of non-terminal ICEs (e.g., treatment dropouts such as treatment discontinuation and rescue medication), but the data after an ICE occurrence may or may not be collected.

Let's read the description of the hypothetical strategy in ICH E9(R1):

> "A scenario is envisaged in which the intercurrent event would not occur: the value of the variable to reflect the clinical question of interest is the value that the variable would have taken in the hypothetical scenario defined."—ICH E9(R1)

By the hypothetical strategy, we envisage a hypothetical scenario in which the ICE would not occur using counterfactual thinking. That is, the hypothetical strategy corresponds to the "C" component of the PROTECT checklist, as presented in Table 4.1. If there is only one type of ICE and let E be the ICE occurrence indicator, then there would be no protocol deviation is equivalent to that the ICE would not occur. Thus, we have

$$Y_i^{z=1,a=1} = Y_i^{z=1,e=0},$$
$$Y_i^{z=0,a=0} = Y_i^{z=0,e=0}.$$

To summarize, if there is only one type of ICE and we want to apply the hypothetical strategy to handle it, we are interested in the following causal estimand,

$$\theta^*_{\text{PP}} = \mathbb{E}(Y^{z=1,e=0} - Y^{z=0,e=0}) = \mathbb{E}(Y^{z=1,a=1} - Y^{z=0,a=0}). \tag{4.4}$$

Estimator

By the hypothetical strategy, we consider the data—even if they may be collected—after the ICE occurrence as irrelevant. Equivalently, by the hypothetical strategy, we consider the data after the ICE occurrence as "missing" data. We have data $\mathcal{O} = \{(X_i, Z_i, E_i, (1-E_i)Y_i), i = 1, \ldots, N\}$, where X_i is a vector of baseline variables, Z_i is treatment assignment, $E_i = 1$ if the ICE occurs, and $(1-E_i)Y_i$ is the observed relevant outcome. We make three assumptions:

1. The consistency assumption:

$$Y = Y^{z=j,e=0} \text{ if } Z = j \text{ and } E = 0, \text{ for } j = 1, 0; \tag{4.5}$$

2. The "missing" at random assumption:

$$(Y^{z=1,e=0}, Y^{z=0,e=0}) \perp\!\!\!\perp E \mid (X, Z); \tag{4.6}$$

3. The positivity assumption:

$$\mathbb{P}(E = 0 \mid Z = j, X = x) > 0, \text{ for } j = 1, 0; x \in \text{supp}(X). \tag{4.7}$$

Under these three assumptions, following the similar arguments in Chapter 3, we construct the following estimator,

$$\widehat{\theta}_{\text{PP}} = \frac{1}{N} \sum_{i=1}^{N} \left[\widehat{Q}_{E=0}(1, X_i) - \widehat{Q}_{E=0}(0, X_i) \right], \tag{4.8}$$

where $\widehat{Q}_{E=0}(Z, X)$ is an estimator of $Q_{E=0}(Z, X) = \mathbb{E}(Y \mid E = 0, Z, X)$.

Missing data handling

Assume that there are missing data, with missing indicator Δ_i. Let E_i be the ICE indicator. Let $\widetilde{\Delta}_i = 1$ if either $E_i = 1$ or $\Delta_i = 1$; let $\widetilde{\Delta}_i = 0$ if $E_i = 0$ and $\Delta_i = 0$. That is, $\widetilde{\Delta}_i$ is the "missing" indicator, which is a composite variable of E_i and Δ_i. Therefore, by the hypothetical strategy, the amount of the "missing" data is larger than or equal to the amount of missing data; that is, $\sum_{i=1}^{N} \widetilde{\Delta}_i/N \geq \sum_{i=1}^{N} \Delta_i/N$.

We have data $\mathcal{O} = \{(X_i, Z_i, \widetilde{\Delta}_i, (1 - \widetilde{\Delta}_i)Y_i), i = 1, \ldots, N\}$. We make the following three assumptions:

1. The consistency assumption:
$$Y = Y^{z=j,e=0} \text{ if } Z = j \text{ and } \widetilde{\Delta} = 0, \text{ for } j = 1, 0; \tag{4.9}$$

2. The "missing" at random assumption:
$$(Y^{z=0,e=0}, Y^{z=1,e=0}) \perp\!\!\!\perp \widetilde{\Delta} | (X, Z); \tag{4.10}$$

3. The positivity assumption:
$$\mathbb{P}(\widetilde{\Delta} = 0 | Z = j, X = x) > 0, \text{ for } j = 1, 0; x \in \text{supp}(X). \tag{4.11}$$

Under these three assumptions, following the same arguments in Chapter 3, we construct the following estimator,

$$\widehat{\theta}_{\text{PP}} = \frac{1}{N} \sum_{i=1}^{N} \left[\widehat{Q}_{\widetilde{\Delta}=0}(1, X_i) - \widehat{Q}_{\widetilde{\Delta}=0}(0, X_i) \right], \tag{4.12}$$

where $\widehat{Q}_{\widetilde{\Delta}=0}(Z, X)$ is an estimator of $Q_{\widetilde{\Delta}=0}(Z, X) = \mathbb{E}(Y | \widetilde{\Delta} = 0, Z, X)$.

Figure 4.2 illustrates the hypothetical strategy. By the hypothetical strategy, the "missing" rate is larger than or equal to the missing rate.

4.1.3 The composite variable strategy

Estimand

Assume that there is only one category of ICEs, which are either non-terminal (e.g., rescue medication) or terminal (e.g., death), and the data after an ICE occurrence may or may not be collected.

Let's read ICH E9(R1)'s description of the composite variable strategy:

> "An intercurrent event is considered to be informative about the patient's outcome and is therefore incorporated into the definition of the variable. For example, a patient who discontinues treatment because of

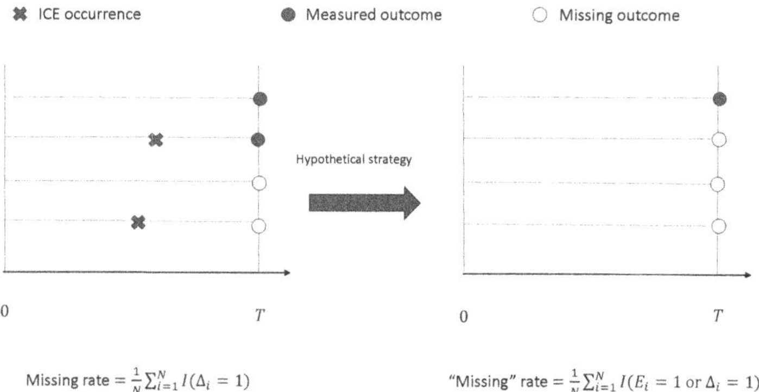

FIGURE 4.2
An illustration of the hypothetical strategy

> toxicity may be considered not to have been successfully treated. If the outcome variable was already success or failure, discontinuation of treatment for toxicity would simply be considered another mode of failure."—ICH E9(R1)

By the composite variable strategy, we define a new outcome variable of interest, which is a composite variable of the original outcome variable and the ICE occurrence. Thus, the composite variable strategy corresponds to the "R/O" component of the PROTECT checklist, as presented in Table 4.1. Define the following new potential outcomes,

$$\breve{Y}^{z=j} = c(Y^{z=j}, E^{z=j}), \tag{4.13}$$

where $c(\cdot, \cdot)$ is a pre-specified composite function of potential outcome $Y^{z=j}$ and potential ICE occurrence $E^{z=j}$, $j = 1, 0$.

Although the composite variable strategy is often used for binary outcomes, ICH E9(R1) provided an example of continuous outcome: "For example, in a trial measuring physical functioning, a variable might be constructed using outcomes on a continuous scale, with subjects who die being attributed a value reflecting the lack of ability to function."

Here are two examples. For binary outcome (with 1 standing for failure and 0 success), we may specify $c(y, e)$ as

$$c_1(y, e) = I(y = 1 \text{ or } e = 1),$$

which equals 1 if $y = 1$ or $e = 1$. For continuous outcome (with a larger value indicating worse outcome), we may specify $c(y, e)$ as

$$c_2(y, e) = y \cdot I(e = 0) + y_{\max} \cdot I(e = 1),$$

where $y_{\max}$ is "a value reflecting the lack of ability to function" .

After the new outcome is defined, the estimand may be defined as

$$\theta^*_{\text{CV}} = \mathbb{E}(\breve{Y}^{z=1} - \breve{Y}^{z=0}). \tag{4.14}$$

Comparing the estimand defined in (4.1) and the one defined in (4.14), we see that the former is the ITT effect on the original outcome variable while the latter is the ITT effect on the newly defined composite outcome variable.

Estimator

Assuming we have observed data $\mathcal{O} = \{(Z_i, \breve{Y}_i), i = 1, \ldots, N\}$, where $\breve{Y}_i = c(Y_i, E_i)$, we can construct the following estimator,

$$\widehat{\theta}_{\text{CV}} = \frac{\sum_{i=1}^N Z_i \breve{Y}_i}{\sum_{i=1}^N Z_i} - \frac{\sum_{i=1}^N (1 - Z_i)\breve{Y}_i}{\sum_{i=1}^N (1 - Z_i)}, \tag{4.15}$$

which is an unbiased estimator for θ^*_{CV} under the consistency assumption:

$$\breve{Y} = \breve{Y}^{z=j} \text{ if } Z = j, \text{ for } j = 1, 0. \tag{4.16}$$

Missing data handling

Assume that there are missing data, with missing indicator Δ_i. Let E_i be the ICE occurrence indicator. Let $\breve{\Delta}_i = 1$ if $E_i = 0$ and $\Delta_i = 1$; let $\breve{\Delta}_i = 0$ if $E_i = 1$ or $\Delta_i = 0$. That is, $\breve{\Delta}_i$ is the "missing" indicator, which is a composite variable of E_i and Δ_i. Unlike the hypothetical strategy that may increase the amount of "missing" data, the composite variable strategy reduces the amount of "missing" data because the ICE occurrence (that may or may not introduce missing data) has been incorporated in the definition of outcome variable $\breve{Y}$. That is, $\sum_{i=1}^N \breve{\Delta}_i/N \leq \sum_{i=1}^N \Delta_i/N$.

Therefore, we have observed data $\mathcal{O} = \{(X_i, Z_i, \breve{\Delta}_i, (1 - \breve{\Delta}_i)\breve{Y}_i), i = 1, \ldots, N\}$. We make the following three assumptions:

1. The consistency assumption:

$$\breve{Y} = \breve{Y}^{z=j} \text{ if } Z = j \text{ and } \breve{\Delta} = 0, \text{ for } j = 1, 0; \tag{4.17}$$

2. The "missing" at random assumption:

$$(\breve{Y}^{z=0}, \breve{Y}^{z=1}) \perp\!\!\!\perp \breve{\Delta} | (X, Z); \tag{4.18}$$

3. The positivity assumption:

$$\mathbb{P}(\breve{\Delta} = 0 | Z = j, X = x) > 0, \text{ for } j = 1, 0; x \in \text{supp}(X). \tag{4.19}$$

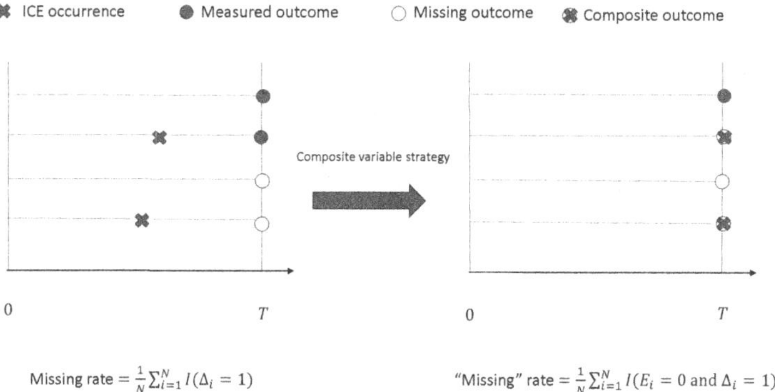

FIGURE 4.3
An illustration of the composite variable strategy

Under these three assumptions, following the same arguments in Chapter 3, we construct the following estimator,

$$\widehat{\theta}_{\mathrm{CV}} = \frac{1}{N}\sum_{i=1}^{N}\left[\widehat{Q}_{\breve{\Delta}=0}(1, X_i) - \widehat{Q}_{\breve{\Delta}=0}(0, X_i)\right], \tag{4.20}$$

where $\widehat{Q}_{\breve{\Delta}=0}(Z, X)$ is an estimator of $Q_{\breve{\Delta}=0}(Z, X) = \mathbb{E}(\breve{Y}|\breve{\Delta} = 0, Z, X)$.

Figure 4.3 illustrates the composite variable strategy. By the composite variable strategy, the "missing" rate is smaller than the missing rate.

4.1.4 The while on treatment strategy

Estimand

Assume that there is only one category of ICEs, which are either non-terminal (e.g., treatment discontinuation) or terminal (e.g., death), and the data after an ICE occurrence may or may not be collected.

Let's read ICH E9(R1)'s description of the while on treatment strategy:

> "For this strategy, response to treatment before the occurrence of the intercurrent event is of interest. Terminology for this strategy will depend on the intercurrent event of interest, e.g., while alive, when considering death as an intercurrent event. If a variable is measured repeatedly, its values up to the time of the intercurrent event may be considered relevant for the clinical question, rather than the value at the same fixed timepoint for all subjects. The same applies to the occurrence of a binary outcome of interest up to the time of the intercurrent event."—ICH E9(R1)

Therefore, by the while on treatment strategy, we revise the "T" component of the PROTECT checklist, as presented in Table 4.1.

To explicitly express the dependence of outcome on the measurement time t, we denote the outcome variable as $Y(t)$. Let T be the primary measurement time, let $\check{T}$ be the last measurement time prior to the occurrence of $E = 1$, with $\check{T}$ being T if the ICE does not occur. In the potential world, let $Y_i^{z=j}(t)$ be the potential outcome at t had the patient been assigned to treatment j, and $\check{T}^{z=j}$ be the time prior to the potential occurrence of the ICE, with $\check{T}^{z=j}$ being T if the ICE would not occur, $j = 1, 0$.

Hence, the estimand may be defined as

$$\theta^*_{\text{WoT}} = \mathbb{E}\{Y^{z=1}(\check{T}^{z=1}) - Y^{z=0}(\check{T}^{z=0})\}. \tag{4.21}$$

Comparing the estimand defined in (4.1) and the one defined in (4.21), we see that the former is the ITT effect on the outcome variable measured at T while the latter is the ITT effect on the outcome variable measured at $\check{T}$—the last measurement time prior to the ICE occurrence.

Estimator

Assuming we have observed data $\mathcal{O} = \{(Z_i, \check{T}_i, Y_i(\check{T}_i)), i = 1, \ldots, N\}$, where $\check{T}_i$ is the observed last measurement time prior to the ICE occurrence with $\check{T}_i$ being T if no ICE occurs, we can construct the following estimator,

$$\widehat{\theta}_{\text{WoT}} = \frac{\sum_{i=1}^N Z_i Y_i(\check{T}_i)}{\sum_{i=1}^N Z_i} - \frac{\sum_{i=1}^N (1 - Z_i) Y_i(\check{T}_i)}{\sum_{i=1}^N (1 - Z_i)}, \tag{4.22}$$

which is an unbiased estimator for θ^*_{WoT} under the consistency assumption:

$$Y(\check{T}) = Y^{z=j}(\check{T}^{z=j}) \text{ if } Z = j, \text{ for } j = 1, 0. \tag{4.23}$$

Missing data handling

Assume that there are missing data, with missing indicator Δ. We can update the definition of $\check{T}$. Let $\check{T}$ be the last measurement time prior to the occurrence of either $E = 1$ or $\Delta = 1$, with $\check{T}$ being T if neither the ICE nor the missing event occurs. Figure 4.4 illustrates the while on treatment strategy. By the while on treatment strategy, there are no "missing" data.

With the updated definition of $\check{T}$, the estimand and the corresponding estimator are the same as the ones in (4.21) and (4.22), respectively.

4.1.5 The principal stratum strategy

Estimand

Assume that there is only one category of ICEs, which are either non-terminal (e.g., infection) or terminal (e.g., death), and the data after an ICE occurrence may or may not be collected.

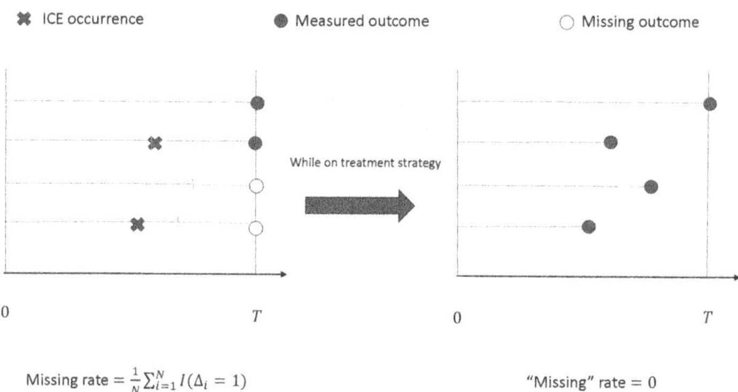

FIGURE 4.4
An illustration of the while on treatment strategy

Let's read the description of the principal stratum strategy in ICH E9(R1):

> "The target population might be taken to be the principal stratum in which an intercurrent event would occur. Alternatively, the target population might be taken to be the principal stratum in which an intercurrent event would not occur. The clinical question of interest relates to the treatment effect only within the principal stratum."—ICH E9(R1)

Therefore, by the principal stratum strategy, we revise the "P" component of the PROTECT checklist, as presented in Table 4.1.

Let E denote the ICE occurrence indicator, where $E = 1$ means an ICE occurs and $E = 0$ means no ICE occurs. In the potential world, let $E^{z=j}$ be the potential occurrence of the ICE had the patient been assigned to treatment j, $j = 1, 0$. There are four principal strata:

$$\begin{aligned} \mathrm{PS}_{11} &= \{i : E_i^{z=1} = 1, E_i^{z=0} = 1\}, \\ \mathrm{PS}_{10} &= \{i : E_i^{z=1} = 1, E_i^{z=0} = 0\}, \\ \mathrm{PS}_{01} &= \{i : E_i^{z=1} = 0, E_i^{z=0} = 1\}, \\ \mathrm{PS}_{00} &= \{i : E_i^{z=1} = 0, E_i^{z=0} = 0\}. \end{aligned}$$

Hence PS_{11} is the principal stratum in which an ICE would occur and PS_{00} is the principal stratum in which an ICE would not occur.

ICH E9(R1) gave two examples. Example 1: "there might be an interest in knowing a treatment effect on severity of infections in the principal stratum of patients becoming infected after vaccination." In this example, E is the infection indicator, and the principal stratum is PS_{11} consisting of patients who would become infected under either treatment assignment 1 or 0. Example 2: "a toxicity might prevent some patients from continuing the test treatment, but there might be an interest in knowing the treatment effect among patients who are able to tolerate the test treatment." In this example, E is the intolerability indicator, and the principal stratum is PS_{00} consisting of patients who would tolerate the treatment under either treatment assignment 1 or 0.

Let $U = (E^{z=1}, E^{z=0})$. Then

$$\mathrm{PS}_{j_1 j_0} = \{U = (j_1, j_0)\}, \ j_1, j_0 = 1, 0. \tag{4.24}$$

By the principal stratum strategy, the estimand may be defined as

$$\theta^*_{\mathrm{PS}_{j_1 j_0}} = \mathbb{E}\{Y^{z=1} - Y^{z=0} | U = (j_1, j_0)\}, \ j_1, j_0 = 1, 0. \tag{4.25}$$

From causal estimand to statistical estimand

The literature on principal stratification is rich; e.g., Frangakis (2002). In this subsection, we follow the discussion of Jiang, Yang, and Ding (2022). We make the following assumptions:

1. The consistency assumption:
$$Y = Y^{z=j} \text{ and } E = E^{z=j} \text{ if } Z = j, \text{ for } j = 1, 0; \tag{4.26}$$

2. The monotonicity assumption:
$$E^{z=0} \leq E^{z=1}; \tag{4.27}$$

3. The principal ignorability assumption:
$$(Y^{z=0}, Y^{z=1}) \perp\!\!\!\perp U | X; \tag{4.28}$$

4. The positivity assumption:
$$\mathbb{P}(U = (j_1, j_0) | X = x) > 0, x \in \mathrm{supp}(X), \tag{4.29}$$
if $\mathrm{PS}_{j_1 j_0}$ is the principal stratum of interest.

Here is a remark on the monotonicity assumption. Under the monotonicity assumption, principal stratum PS_{01} is empty. In the intolerability example, E is the indicator that the patient cannot tolerate the treatment, and therefore the monotonicity assumes that if the patients should not tolerate treatment 0, they would not tolerate treatment 1 either, while if the patients should tolerate treatment 1, they would tolerate treatment 0 as well. In the infection example, E is the indicator that the patient is infected, so we should consider the mirror version of the monotonicity assumption: $E_i^{z=1} \leq E_i^{z=0}$, which implies that, if the vaccinated patients were to be infected, they would be infected had they not been vaccinated, while if the non-vaccinated patients were to be uninfected, they would be uninfected had they been vaccinated.

Table 4.2 shows the relationship between the observed strata defined by (Z, E) and the principal strata under the monotonicity assumption. First, consider the observed stratum $(Z, E) = (0, 0)$. By the consistency assumption,

TABLE 4.2
Principal strata in the observed strata under monotonicity

	$E=0$	$E=1$
$Z=0$	$U \in \{(0,0),(1,0)\}$	$U=(1,1)$
$Z=1$	$U=(0,0)$	$U \in \{(1,1),(1,0)\}$

$E^{z=0}=0$. Moreover, by the monotonicity assumption, $E^{z=1} \geq E^{z=0}=0$; that is, $E^{z=1}=1$ or $E^{z=1}=0$. Therefore, $(Z,E)=(0,0)$ consists of two principal strata, PS_{10} and PS_{00}. Second, consider the observed stratum $(Z,E)=(0,1)$. By the consistency assumption, $E^{z=0}=1$. Moreover, by the monotonicity assumption $E^{z=1} \geq E^{z=0}=1$; that is, $E^{z=1}=1$. Therefore, $(Z,E)=(0,1)$ is the same as the principal stratum PS_{11}. Similarly, we can verify that the observed stratum $(Z,E)=(1,0)$ is the same as the principal stratum PS_{00} and that $(Z,E)=(1,1)$ consists of two principal strata, PS_{11} and PS_{10}.

Here is a remark on the principal ignorability assumption. Jiang, Yang, and Ding (2022) considered a weaker version of the principal ignorability assumption: $\mathbb{E}\{Y^{z=1}|U=(1,1),X\}=\mathbb{E}\{Y^{z=1}|U=(1,0),X\}$ and $\mathbb{E}\{Y^{z=0}|U=(0,0),X\}=\mathbb{E}\{Y^{z=0}|U=(0,1),X\}$. Here we consider the stronger version because its expression looks similar to the MAR assumption.

Assume PS_{11} is the principal stratum of interest. We now show how to translate $\theta^*_{\mathrm{PS}_{11}}=\mathbb{E}[Y^{z=1}-Y^{z=0}|U=(1,1)]$ into a statistical estimand. Note

$$\theta^*_{\mathrm{PS}_{11}}=\mathbb{E}[Y^{z=1}|U=(1,1)]-\mathbb{E}[Y^{z=0}|U=(1,1)]. \tag{4.30}$$

The first term on the right-hand-side (RHS) of (4.30) is

$$\begin{aligned}
&\mathbb{E}[Y^{z=1}|U=(1,1)]\\
\overset{(a)}{=}&\mathbb{E}\{\mathbb{E}[Y^{z=1}|U=(1,1),X]|U=(1,1)\}\\
\overset{(b)}{=}&\mathbb{E}\{\mathbb{E}[Y^{z=1}|U=(1,1)\text{ or }(1,0),X]|U=(1,1)\}\\
\overset{(c)}{=}&\mathbb{E}\{\mathbb{E}[Y^{z=1}|Z=1,U=(1,1)\text{ or }(1,0),X]|U=(1,1)\}\\
\overset{(d)}{=}&\mathbb{E}\{\mathbb{E}[Y|Z=1,U=(1,1)\text{ or }(1,0),X]|U=(1,1)\}\\
\overset{(e)}{=}&\mathbb{E}\{\mathbb{E}[Y|Z=1,E=1,X]|U=(1,1)\}\\
\overset{(f)}{=}&\mathbb{E}\{Q_{11}(X)|U=(1,1)\}\\
\overset{(g)}{=}&\int Q_{11}(x)d\mathcal{F}(x|U=(1,1))\\
\overset{(h)}{=}&\int Q_{11}(x)\frac{\mathbb{P}(U=(1,1)|x)d\mathcal{F}(x)}{\mathbb{P}(U=(1,1))}\\
\overset{(i)}{=}&\mathbb{E}\left\{\frac{\mathbb{P}(U=(1,1)|X)}{\mathbb{P}(U=(1,1))}Q_{11}(X)\right\},
\end{aligned}$$

where (a) holds using the law of iterated expectations, (b) holds under the principal ignorability assumption, (c) holds because of the randomization (either complete randomization or stratified randomization if X includes the stratification factor S), (d) holds under the consistency assumption, (e) holds under the monotonicity assumption with the help of Table 4.2, in (f)

$$Q_{ze}(X) = \mathbb{E}(Y|Z = z, E = e, X), \text{ for } z, e = 0, 1, \tag{4.31}$$

are the subgroup-specific regression functions of Y against X, in (g) the expectation is written as integration with probability distribution $\mathcal{F}$, (h) holds using Bayes's formula, and in (i) the integration is written as expectation.

Similarly, the second term on RHS of (4.30) is

$$\begin{aligned}
&\mathbb{E}[Y^{z=0}|U = (1,1)] \\
=&\mathbb{E}\{\mathbb{E}[Y^{z=0}|U = (1,1), X]|U = (1,1)\} \\
=&\mathbb{E}\{\mathbb{E}[Y^{z=0}|Z = 0, U = (1,1), X]|U = (1,1)\} \\
=&\mathbb{E}\{\mathbb{E}[Y|Z = 0, U = (1,1), X]|U = (1,1)\} \\
=&\mathbb{E}\{\mathbb{E}[Y|Z = 0, E = 1, X]|U = (1,1)\} \\
=&\mathbb{E}\{Q_{01}(X)|U = (1,1)\} \\
=&\int Q_{01}(x) d\mathcal{F}(x|U = (1,1)) \\
=&\int Q_{01}(x) \frac{\mathbb{P}(U = (1,1)|x) d\mathcal{F}(x)}{\mathbb{P}(U = (1,1))} \\
=&\mathbb{E}\left\{\frac{\mathbb{P}(U = (1,1)|X)}{\mathbb{P}(U = (1,1))} Q_{01}(X)\right\}.
\end{aligned}$$

Combining the above two results, we have

$$\theta^*_{\text{PS}_{11}} = \mathbb{E}\left\{\frac{\mathbb{P}(U = (1,1)|X)}{\mathbb{P}(U = (1,1))} [Q_{11}(X) - Q_{01}(X)]\right\}. \tag{4.32}$$

Similarly, we can show that

$$\theta^*_{\text{PS}_{00}} = \mathbb{E}\left\{\frac{\mathbb{P}(U = (0,0)|X)}{\mathbb{P}(U = (0,0))} [Q_{10}(X) - Q_{00}(X)]\right\}, \tag{4.33}$$

$$\theta^*_{\text{PS}_{10}} = \mathbb{E}\left\{\frac{\mathbb{P}(U = (1,0)|X)}{\mathbb{P}(U = (1,0))} [Q_{11}(X) - Q_{00}(X)]\right\}. \tag{4.34}$$

However, (4.32)–(4.34) are not statistical estimands, because they are still dependent on U. Since they are not statistical estimands, we need to continue the identification process until we get rid of U. For this aim, define the following principal scores (Ding and Lu 2017),

$$u_{j_1 j_0}(X) = \mathbb{P}(U = (j_1, j_0)|X), \; j_1, j_0 = 1, 0, \tag{4.35}$$

which are equal to the proportions of the principal strata given the covariates. In the following, we attempt to express the principal scores in terms of observed data so that we can complete the task of identification.

Assume we have observed data $\{(X_i, Z_i, E_i, Y_i), i = 1, \ldots, N\}$. Define the following ICE occurrence probability conditional on X and Z,

$$e_j(X) = \mathbb{P}(E = 1|Z = j, X), \; j = 1, 0. \tag{4.36}$$

We can show that

$$\begin{aligned} u_{11}(X) &= \mathbb{P}(U = (1,1)|X) \\ &\overset{(a)}{=} \mathbb{P}(U = (1,1)|Z = 0, X) \\ &\overset{(b)}{=} \mathbb{P}(E = 1|Z = 0, X) \\ &\overset{(c)}{=} e_0(X), \end{aligned}$$

where (a) holds because of the randomization (either complete randomization or stratified randomization if covariates X include stratification factor S), (b) holds under the monotonicity assumption with the help of Table 4.2, and (c) uses the definition in (4.36).

Similarly, we can show that

$$\begin{aligned} u_{00}(X) &= \mathbb{P}(U = (0,0)|X) \\ &= \mathbb{P}(U = (0,0)|Z = 1, X) \\ &= \mathbb{P}(E = 1|Z = 1, X) \\ &= 1 - e_1(X). \end{aligned}$$

In addition, since $u_{11}(X) + u_{00}(X) + u_{10}(X) = 1$, we have

$$u_{10}(X) = e_1(X) - e_0(X). \tag{4.37}$$

Let

$$\begin{aligned} e_j &= \mathbb{E}\{e_j(X)\}, \; j = 1, 0; \\ u_{j_1 j_0} &= \mathbb{E}\{u_{j_1 j_0}(X)\}, \; j_1, j_0 = 1, 0. \end{aligned}$$

Then we have

$$\begin{aligned} u_{11} &= e_0, \\ u_{00} &= 1 - e_1, \\ u_{10} &= e_1 - e_0. \end{aligned}$$

Using these results, we complete the task of translating causal estimands into statistical estimands:

$$\theta^*_{\mathrm{PS}_{11}} = \mathbb{E}\left\{\frac{e_0(X)}{e_0}\left[Q_{11}(X) - Q_{01}(X)\right]\right\} \triangleq \theta_{\mathrm{PS}_{11}}, \tag{4.38}$$

$$\theta^*_{\mathrm{PS}_{00}} = \mathbb{E}\left\{\frac{1 - e_1(X)}{1 - e_1}\left[Q_{10}(X) - Q_{00}(X)\right]\right\} \triangleq \theta_{\mathrm{PS}_{00}}, \tag{4.39}$$

$$\theta^*_{\mathrm{PS}_{10}} = \mathbb{E}\left\{\frac{e_1(X) - e_0(X)}{e_1 - e_0}\left[Q_{11}(X) - Q_{00}(X)\right]\right\} \triangleq \theta_{\mathrm{PS}_{10}}. \tag{4.40}$$

Estimator

Consider the statistical estimands defined in (4.38)–(4.40). If we can obtain estimators for $e_j(X)$, denoted as $\widehat{e}_j(X)$ (say, using logistics regression), and obtain estimators for $Q_{ze}(X)$, denoted as $\widehat{Q}_{ze}(X)$ (say, using linear regression for continuous outcome and logistics regression for binary outcome), then we can obtain the following estimators,

$$\widehat{\theta}_{\text{PS}_{11}} = \frac{1}{N}\sum_{i=1}^{N} \frac{\widehat{e}_0(X_i)}{\widehat{e}_0} \left[\widehat{Q}_{11}(X_i) - \widehat{Q}_{01}(X_i)\right], \tag{4.41}$$

$$\widehat{\theta}_{\text{PS}_{00}} = \frac{1}{N}\sum_{i=1}^{N} \frac{1-\widehat{e}_1(X_i)}{1-\widehat{e}_1} \left[\widehat{Q}_{10}(X_i) - \widehat{Q}_{00}(X_i)\right], \tag{4.42}$$

$$\widehat{\theta}_{\text{PS}_{10}} = \frac{1}{N}\sum_{i=1}^{N} \frac{\widehat{e}_1(X_i)-\widehat{e}_0(X_i)}{\widehat{e}_1-\widehat{e}_0} \left[\widehat{Q}_{11}(X_i) - \widehat{Q}_{00}(X_i)\right], \tag{4.43}$$

where $\widehat{e}_j = \sum_{i=1}^{N} \widehat{e}(X_i)/N$, $j = 1, 0$.

Missing data handling

There may be missing data on outcome Y_i and there may be missing data on ICE E_i at the same time. For the purpose of demonstration, in this subsection, we consider the scenario where there are only missing data on outcome Y_i, with the missing indicator denoted as Δ_i. We have observed data $\mathcal{O} = \{(X_i, Z_i, E_i, \Delta_i, (1-\Delta_i)Y_i), i = 1, \ldots, N\}$. We also assume the missing at random assumption and define the following regression functions,

$$Q_{ze|\Delta=0}(X) = \mathbb{E}(Y|\Delta = 0, Z = z, E = e, X), \text{ for } z, e = 0, 1. \tag{4.44}$$

If we can obtain estimators for $Q_{ze|\Delta=0}(X)$, denoted as $\widehat{Q}_{ze|\Delta=0}(X)$ (say, using linear regression for continuous outcome and logistics regression for binary outcome), then we can obtain the following estimators,

$$\widehat{\theta}_{\text{PS}_{11}} = \frac{1}{N}\sum_{i=1}^{N} \frac{\widehat{e}_0(X_i)}{\widehat{e}_0} \left[\widehat{Q}_{11|\Delta=0}(X_i) - \widehat{Q}_{01|\Delta=0}(X_i)\right], \tag{4.45}$$

$$\widehat{\theta}_{\text{PS}_{00}} = \frac{1}{N}\sum_{i=1}^{N} \frac{1-\widehat{e}_1(X_i)}{1-\widehat{e}_1} \left[\widehat{Q}_{10|\Delta=0}(X_i) - \widehat{Q}_{00|\Delta=0}(X_i)\right], \tag{4.46}$$

$$\widehat{\theta}_{\text{PS}_{10}} = \frac{1}{N}\sum_{i=1}^{N} \frac{\widehat{e}_1(X_i)-\widehat{e}_0(X_i)}{\widehat{e}_1-\widehat{e}_0} \left[\widehat{Q}_{11|\Delta=0}(X_i) - \widehat{Q}_{00|\Delta=0}(X_i)\right]. \tag{4.47}$$

4.2 Combinations of Strategies

In practice, we may handle different types of ICEs using different strategies. If there is only one type of ICEs, there are five strategies. If there are two types of ICEs that we want to handle differently, there are $5 \times 4 = 20$ combinations of two different strategies. If there are three types of ICEs that we want to handle differently, there are $5 \times 4 \times 3 = 60$ combinations of three different strategies. But of course, some combinations have been less common than others, rarely been used, or never been used. We provide some examples of commonly used combinations.

Example 1: Combination of the treatment policy strategy and the hypothetical strategy

Assume there are two types of ICEs, denoted as $E(1)$ (e.g., treatment discontinuation without alternative treatment) and $E(2)$ (e.g., rescue medication), along with analysis dropout denoted as Δ.

Method 1: We consider the treatment policy strategy to handle both $E(1)$ and $E(2)$. But then how to handle analysis dropout Δ? There are two approaches for handling Δ. One approach is to consider the problem of handling Δ as a missing data problem and then relegate the discussion of missing data handling to the estimation stage. The other approach is to consider Δ as another type of ICE and handle it using the hypothetical strategy—envisaging a hypothetical scenario where the analysis dropout would not occur.

In theory, these two approaches are equivalent under the missing at random assumption. But the two approaches are fundamentally different. The former approach doesn't explicitly specify the strategy for handling Δ in the stage of estimand defining, while the latter approach does. To make the estimand definition crystal clear, we recommend the latter approach and define the following estimand:

$$\theta^* = \mathbb{E}(Y^{z=1,\delta=0} - Y^{z=0,\delta=0}), \tag{4.48}$$

where $Y^{z=j,\delta=0}$ is the potential outcome if the subject had been assigned treatment j, $j = 1, 0$, and the analysis dropout would not occur. Both potential ICE occurrences, $E^{z=j}(1)$ and $E^{z=j}(2)$, are omitted from the definition of this estimand because they are handled by the treatment policy strategy.

Method 2: We consider using the treatment policy strategy to handle either $E(1)$ or $E(2)$ (say, $E(1)$), and using the hypothetical strategy to handle the other one (say, $E(2)$). Again, how to handle analysis dropout Δ? There are two approaches to handling Δ. Following the above discussion, we take the approach of handling analysis dropout Δ via the hypothetical strategy—envisaging a hypothetical scenario where the analysis dropout would not occur. Since both $E(2)$ and Δ are to be handled by the hypothetical strategy, we group the two into $\widetilde{\Delta}$, where $\widetilde{\Delta} = 1$ if $E(2) = 1$ or $\Delta = 1$. To make the estimand definition crystal clear, we define the following estimand:

$$\theta^* = \mathbb{E}(Y^{z=1,\widetilde{\delta}=0} - Y^{z=0,\widetilde{\delta}=0}), \tag{4.49}$$

where $Y^{z=j,\widetilde{\delta}=0}$ is the potential outcome if the subject had been assigned treatment j, $j = 1, 0$, and the composite ICE, $\widetilde{\Delta}$, would not occur. Potential ICE occurrences of $E(1)$, $E^{z=j}(1)$, $j = 1, 0$, are omitted from the definition of this estimand because $E(1)$ is handled by the treatment policy strategy.

Method 3: We consider using the hypothetical strategy to handle both $E(1)$ and $E(2)$. Again, how to handle analysis dropout Δ? There are two approaches for handling Δ. Following the above discussion, we take the approach of handling analysis dropout Δ via the hypothetical strategy—envisaging a hypothetical scenario where the analysis dropout would not occur. Since we want to handle $E(1)$, $E(2)$, and Δ by the hypothetical strategy, we group them into $\widetilde{\Delta}$, where $\widetilde{\Delta} = 1$ if $E(1) = 1$, $E(2) = 1$, or $\Delta = 1$. To make the estimand definition crystal clear, we define the following estimand:

$$\theta^* = \mathbb{E}(Y^{z=1,\widetilde{\delta}=0} - Y^{z=0,\widetilde{\delta}=0}) = \mathbb{E}(Y^{z=1,a=1} - Y^{z=0,a=0}), \tag{4.50}$$

where $Y^{z=j,\widetilde{\delta}=0}$ is the potential outcome if the subject had been assigned to treatment j, $j = 1, 0$, and the composite ICE, $\widetilde{\Delta}$, would not occur.

Example 2: Combination of the treatment policy strategy, the hypothetical strategy, and the composite variable strategy

Assume there are three types of ICEs, denoted as $E(1)$ (e.g., treatment discontinuation without alternative treatment), $E(2)$ (e.g., rescue medication), $E(3)$ (e.g., death), along with analysis dropout denoted as Δ. Assume that outcome Y is binary, with 1 standing for failure and 0 standing for success.

Method 1: We consider the treatment policy strategy to handle both $E(1)$ and $E(2)$, the composite variable strategy to handle $E(3)$, and the hypothetical strategy to handle Δ. Since we use the composite variable strategy to handle $E(3)$, we define the new outcome variable $\widetilde{Y}$, where $\widetilde{Y} = 1$ if $E(3) = 1$ or $Y = 1$. Then we define the following estimand:

$$\theta^* = \mathbb{E}(\widetilde{Y}^{z=1,\delta=0} - \widetilde{Y}^{z=0,\delta=0}). \tag{4.51}$$

Method 2: We consider the treatment policy strategy to handle $E(1)$, the hypothetical strategy to handle $E(2)$, the composite variable strategy to handle $E(3)$, and the hypothetical strategy to handle Δ. We define new composite outcome $\widetilde{Y}$, where $\widetilde{Y} = 1$ if $E(3) = 1$ or $Y = 1$. We define the new composite ICE $\widetilde{\Delta}$, where $\widetilde{\Delta} = 1$ if $E(2) = 1$ or $\Delta = 1$. Then we define the following estimand:

$$\theta^* = \mathbb{E}(\widetilde{Y}^{z=1,\widetilde{\delta}=0} - \widetilde{Y}^{z=0,\widetilde{\delta}=0}). \tag{4.52}$$

Method 3: We consider the hypothetical strategy to handle $E(1)$, and the composite variable strategy to handle $E(2)$, $E(3)$, and Δ. We define new composite outcome $\widetilde{Y}$, where $\widetilde{Y} = 1$ if $E(2) = 1$, $E(3) = 1$, or $\Delta = 1$. Then we define the following estimand:

$$\theta^* = \mathbb{E}(\widetilde{Y}^{z=1,e(1)=0} - \widetilde{Y}^{z=0,e(1)=0}), \tag{4.53}$$

where $\widetilde{Y}^{z=j,e(1)=0}$ is the potential outcome if the subject had been assigned to treatment j, $j = 1, 0$, and ICE $E(1)$ would not occur.

Method 4: We consider the composite variable strategy to handle $E(1)$, $E(2)$, $E(3)$, and Δ. We define new composite outcome $\widetilde{Y}$, where $\widetilde{Y} = 1$ if $E(1) = 1$, $E(2) = 1$, $E(3) = 1$, or $\Delta = 1$. Then we define the following estimand:

$$\theta^* = \mathbb{E}(\widetilde{Y}^{z=1} - \widetilde{Y}^{z=0}). \tag{4.54}$$

4.3 Time-to-event Outcome

So far we have focused on continuous outcomes and binary outcomes. In this section, we discuss how to apply the above strategies to handle ICEs in clinical trials with time-to-event outcomes. This chapter is already too long, so we will take two steps to discuss the topic of time-to-event outcome. In this section, we will briefly discuss the strategies for ICE handling in the estimand construction, leaving detailed discussion to the next chapter.

4.3.1 Censoring

For time-to-event outcomes, there are several types of events. First, we refer to the event in the definition of the time-to-event outcome as a primary event (PE). Next, there may be ICEs, as we discuss in the previous section. In addition, different from continuous or binary outcomes, for time-to-event outcomes, there is the issue of right censoring. Right censoring occurs when a participant leaves the study before the PE occurs (e.g., a consequence of some ICE) or the study ends before the PE occurs (i.e., administrative censoring). We refer the event of right censoring as a censoring event (CE).

To supply some mathematical symbols to the times when these events occur. We denote the time to PE as Y, which is the primary outcome. Also, denote the time to CE as C. Because of potential right censoring, we observe the minimum of Y and C, denoted by $Y^* = \min(Y, C)$. Let $\Delta = I(C < Y)$ be the censoring status, with $\Delta = 1$ standing for being censored.

Let $Y^{z=j,\delta=0}$ be the potential outcome had the patient been assigned treatment j and not censored, $j = 1, 0$. Then we may be interested in the following estimand in terms of restricted mean survival time (RMST),

$$\theta_1^* = \mathbb{E}\left\{\min(Y^{z=1,\delta=0}, T) - \min(Y^{z=0,\delta=0}, T)\right\}, \tag{4.55}$$

or the following estimand in terms of survival rate,

$$\theta_2^* = \mathbb{E}\left\{I(Y^{z=1,\delta=0} > T) - I(Y^{z=0,\delta=0} > T)\right\}. \tag{4.56}$$

Note that in the above definitions of θ_1^* and θ_2^*, we implicitly apply the hypothetical strategy to handle the CE, which is treated as an ICE, by envisaging a hypothetical scenario where the CE would not occur.

4.3.2 The treatment policy strategy

Assume that there is one non-terminal ICE and the data after an ICE are still collected. If we apply the treatment policy strategy to handle such ICE, we revise the "T/E" component of the PROTECT checklist, comparing two treatment policies instead of comparing two treatments. The estimands under this strategy are the same as those defined in (4.55) and (4.56).

4.3.3 The hypothetical strategy

Assume that there is one non-terminal ICE (e.g., treatment discontinuation or rescue medication), regardless of whether the data after the ICE occurrence are collected. If we apply the hypothetical strategy, we revise the "C" component of the PROTECT checklist, by envisaging a hypothetical scenario where the ICE would not occur. Define the ICE occurrence indicator as E. After we define a new "censoring" event indicator, $\widetilde{\Delta} = 1$ if $E = 1$ or $\Delta = 1$, the estimands under this strategy are the same as those defined in (4.55) and (4.56) except that $\delta = 0$ is replaced by $\widetilde{\delta} = 0$.

4.3.4 The composite variable strategy

Assume that there is one ICE, which is either terminal or non-terminal. If we apply the composite variable strategy, we revise the "R/O" component of the PROTECT checklist, by defining a new time-to-event outcome, $\widetilde{Y}$, which is the time for ICE or PE, whichever occurs first. After we define the new outcome variable $\widetilde{Y}$, the estimands under this strategy are the same as those defined in (4.55) and (4.56) except that $Y^{z=j,\delta=0}$ is replaced by $\widetilde{Y}^{z=j,\delta=0}$.

4.3.5 The while on treatment strategy

Assume that there is one ICE, which is either terminal or non-terminal. To see if we could apply the while on treatment strategy, we read the description of the while on treatment strategy in ICH E9(R1) one more time: "For this strategy, response to treatment before the occurrence of the intercurrent event is of interest." This strategy works well for continuous and binary outcomes, as we see in the previous section because they are measurable at the time point immediately prior to the ICE occurrence. However, it is hard to apply this strategy for time-to-event outcome, which is not measurable prior to the ICE occurrence—we know for sure the PE hasn't occurred when the ICE occurs but we don't know anything about when the PE will occur at the time when the ICE occurs.

4.3.6 The principal stratum strategy

Assume that there is one ICE, either terminal or non-terminal. If we apply the principal stratum strategy, we revise the "P" component of the PROTECT checklist, by defining a principal stratum of interest. Using the same notation in the previous section, we select one principal stratum, PS_{11}, PS_{10}, PS_{01}, or PS_{00}. After we determine the principal stratum of interest, the estimands under this strategy are the same as those defined in (4.55) and (4.56) except that the expectation is with respect to the principal stratum of interest.

4.3.7 The competing risk strategy

A competing risk (CR) is one ICE that competes with the PE, in the sense that the occurrence of one CR event prevents the occurrence of the PE and vice versa. For example, in a study examining time to death attributable to lung cancer, death attributable to other causes is CR. Therefore, we propose to add the competing-risk strategy to the toolbox that already contains the five strategies proposed by ICH E9(R1).

The literature on competing risk analysis is rich. See, for example, Prentice et al. (1978) and Austin, Lee, and Fine (2016). A common population-level summary to evaluate the treatment effect is the cumulative incidence function (CIF). Denote the time to CR as R, with two potential outcomes, $R^{z=1}$ and $R^{z=0}$. Then we may define the following estimand in terms of CIF,

$$\theta^* = \mathbb{E}\left\{I(Y^{z=1} < T, Y^{z=1} < R^{z=1}) - I(Y^{z=0} < T, Y^{z=0} < R^{z=0})\right\}. \quad (4.57)$$

4.4 Sample Size Calculation

When planning clinical trials with potential missing data, a conventional approach is two-stage:

1. Without considering the impact of missing data or ICEs, we calculate a required sample size N;
2. Assuming an expected missing data rate r, we adjust the sample size from N to $N^* = N/(1-r)$.

However, this conventional approach should be used with caution for at least three reasons. First, this approach is not estimand oriented. Sample size calculation is a key step in clinical trial planning, so it should be aligned with the estimand. Second, since ICEs are sources of some missing data or "missing" data when we conduct sample size calculation, we should adjust for the impact of ICEs. Third, the amount of missing data or "missing" data depends on the strategies selected to handle ICEs. For example, the amount of missing data in an analysis aligned with the treatment policy strategy will be smaller than the amount of "missing" data in an analysis aligned with the hypothetical strategy.

Fang and Jin (2021) proposed five basic approaches to aligning sample size calculation with the estimand, with each approach corresponding to one of the five ICE handling strategies. In this section, we briefly describe these five basic approaches, using binary outcomes as an example. The approaches can be easily extended to continuous outcomes (Fang and Jin, 2021) and time-to-event outcomes (Fang, Jin, and Wu 2024).

Consider a two-arm RCT. Let Y be a binary outcome measured at T. In a perfect RCT, the effect of interest is

$$\theta^*_{\text{ITT}} = \mathbb{P}(Y^{z=1} = 1) - \mathbb{P}(Y^{z=0} = 1), \tag{4.58}$$

which is the same as

$$\theta^*_{\text{PP}} = \mathbb{P}(Y^{z=1,a=1} = 1) - \mathbb{P}(Y^{z=0,a=0} = 1). \tag{4.59}$$

Example 0: Assume we are planning a 1:1 RCT with a binary outcome to detect the difference between the expected response proportion $p_1 = 0.5$ in the treatment arm and the expected response proportion $p_0 = 0.3$ in the control arm. The expected treatment effect is $\theta = p_1 - p_0$ and the effect size is $ES = |\theta|/\sqrt{p(1-p)}$, where $p = (p_1 + p_0)/2$. Assume we want to achieve $1 - \beta = 80\%$ power under significance level $\alpha = 5\%$. Under the ideal scenario where there are no missing data or ICEs, based on the following formula—see e.g. Chow et al. (2017), where Z_α denotes the $100\alpha\%$ of a standard normal distribution,

$$N = 2 \times \left[2 \left(\frac{Z_{1-\alpha/2} + Z_{1-\beta}}{ES} \right)^2 \right], \tag{4.60}$$

the required sample size is $N = 2 \times 94 = 188$ to two arms combined.

4.4.1 The treatment policy strategy

Assume that there is only one ICE to be handled by the treatment policy strategy. Assume that the proportion of subjects who are expected to have ICE in arm $Z = j$ is r_j, $j = 1, 0$. To be conservative, we assume that, in the treatment arm, the distribution of outcome Y after the occurrence of an ICE is the same as that in the control control arm. Under this assumption, we have

$$\begin{aligned} \mathbb{P}(Y^{z=1} = 1) &= (1 - r_1)\mathbb{P}(Y^{z=1,a=1} = 1) + r_1\mathbb{P}(Y^{z=1,a=0} = 1) \\ &= (1 - r_1)\mathbb{P}(Y^{z=1,a=1} = 1) + r_1\mathbb{P}(Y^{z=0} = 1). \end{aligned}$$

Example 1: We continue the discussion of Example 0. Furthermore, we assume that $r_1 = 15\%$ and $r_0 = 10\%$. By the treatment policy strategy, the original comparison of $p_1 = 0.5$ vs. $p_0 = 0.3$ is diluted as the comparison of $p_1' = 0.5(1-0.15)+0.3(0.15) = 0.47$ vs. $p_0' = 0.3$. For this diluted comparison, the required sample size is $N^* = 2 \times 128 = 256$.

4.4.2 The hypothetical strategy

Assume that there is only one ICE to be handled by the hypothetical strategy. Assume that the proportion of subjects who are expected to have ICE in arm $Z = j$ is r_j, $j = 1, 0$. By the hypothetical strategy, the estimand of interest is θ^*_{PP} and the proportion of ICE in each arm is the proportion of missing data in each arm. Thus, the conventional approach is appropriate.

Example 2: We continue the discussion of Example 0. Furthermore, we assume that $r_1 = 15\%$ and $r_0 = 10\%$. Without adjusting for missing data or ICEs, the required sample size is $N = 188$. Using the conventional approach, the required sample size is adjusted as $N^* = N/[1-(r_1+r_0)/2] = 188/(1-0.125) = 216$. Note that the sample size is rounded up to an even number such that it can be divided into two arms.

4.4.3 The composite variable strategy

Assume that there is only one ICE to be handled by the composite variable strategy. Assume that the proportion of subjects who are expected to have ICE in arm $Z = j$ is r_j, $j = 1, 0$. If Y is binary (say, 1 stands for failure and 0 stands for success), define a new outcome variable $\widetilde{Y}$, such as $\widetilde{Y} = 1$ if $Y = 1$ or an ICE occurs. Then, we have

$$\mathbb{P}(\widetilde{Y}^{z=j} = 1) = (1 - r_j)\mathbb{P}(Y^{z=j} = 1) + r_j, j = 1, 0.$$

Example 3: We continue the discussion of Example 0. In this example, Y is binary with 1 standing for failure, so the comparison becomes $q_1 = 1-p_1 = 0.5$ vs. $q_0 = 1 - p_0 = 0.7$. Furthermore, we assume that $r_1 = 15\%$ and $r_0 = 10\%$. By the composite variable strategy, the original comparison of $q_1 = 0.5$ vs. $q_0 = 0.7$ becomes the new comparison of $q_1' = [0.5(1-0.15)+0.15] = 0.575$ vs. $q_0' = 0.7(1-0.1)+0.1 = 0.73$. For the new comparison, the required sample size is $N^* = 2 \times 147 = 294$.

4.4.4 The while on treatment strategy

Assume that there is only one ICE to be handled by the while on treatment strategy. Assume that the proportion of subjects who are expected to have ICE in arm $Z = j$ is r_j, $j = 1, 0$. By the while on treatment strategy, we define the new outcome variable $\widetilde{Y}$ as the outcome variable measured at the time immediately prior to the ICE occurrence if an ICE occurs. Furthermore, assume that, in the treatment arm, the treatment effect at a given time is proportional to the treatment duration up to that time, while in the control arm, the treatment effect is the same regardless of the ICE occurrence. Moreover, assume the time of ICE occurrence is following a uniform distribution between 0 and T. Under these assumptions, we have

$$\mathbb{P}(\widetilde{Y}^{z=1} = 1) = (1 - r_1)\mathbb{P}(Y^{z=1} = 1) + r_1 \left[\mathbb{P}(Y^{z=1} = 1) + \mathbb{P}(Y^{z=0} = 1)\right]/2.$$

Example 4: We continue the discussion of Example 0. Furthermore, we assume that $r_1 = 15\%$ and $r_0 = 10\%$. By the while on treatment strategy, the original comparison of $p_1 = 0.5$ vs. $p_0 = 0.3$ is diluted as the new comparison of $p_1' = (1-0.15)(0.5)+0.15(0.5+0.3)/2 = 0.485$ vs. $p_0' = 0.3$. For this diluted comparison, the required sample size is $N^* = 2 \times 109 = 218$.

4.4.5 The principal stratum strategy

Assume that there is only one ICE to be handled by the principal stratum strategy. Assume that the proportion of subjects who are expected to have ICE in arm $Z = j$ is r_j, $j = 1, 0$. Assume we are interested in the principal stratum $\text{PS}_{00} = \{E^{z=1} = 0, E^{z=0} = 0\}$. We have

$$\begin{aligned}\mathbb{P}(E^{z=1} = 0, E^{z=0} = 0) &= 1 - \mathbb{P}(E^{z=1} = 1 \text{ or } E^{z=0} = 1)\\ &\geq 1 - [\mathbb{P}(E^{z=1} = 1) + \mathbb{P}(E^{z=0} = 1)]\\ &= 1 - (r_1 + r_0).\end{aligned}$$

Thus, the required sample size is $N^* = N/(1 - r_1 - r_0)$.

Example 5: We continue the discussion of Example 0. Furthermore, we assume that $r_1 = 15\%$ and $r_0 = 10\%$. Thus, if the principal stratum PS_{00} is of interest, the needed sample size is $N^* = N/(1-0.15-0.1) = 188/0.75 = 252$.

4.5 Exercises

In R, generate a data frame that has 5 columns (subject ID `SID`, covariate `X`, treatment assignment `Z` being 0 or 1, ICE indicator `E` being 0 or 1, and observed incomplete outcome `Yobs`) and 100 rows, using the following R codes:

```
set.seed (6)
SID <- 1:100
X <- rnorm(n=100, mean=0, sd=2)
Z <- rbinom(n=100, size=1, prob=0.5)

E1 <- rbinom (n=100, size=1, prob=exp(-2.5+X)/(1+exp(-2.5+X)))
E0 <- rbinom (n=100, size=1, prob=exp(-3.0+X)/(1+exp(-3.0+X)))

E <- Z*E1 + (1-Z)*E0
```

```
mean(E) # ICE rate

Y00<-rnorm(n=100, mean=0, sd=1)
Y01<-rnorm(n=100, mean=-0.3, sd=1)
Y11<-rnorm(n=100, mean=1, sd=1)
Y10<-rnorm(n=100, mean=0.5, sd=1)

Yobs <- (1-Z)*(1-E)*Y00+(1-Z)*E*Y01+Z*(1-E)*Y10+Z*E*Y11
summary(Yobs) # summary of observed outcome

dataset6 <- data.frame(SID=SID, X=X, Z=Z, E=E, Y=Yobs)
```

Note that in the above population, E1 is $E^{z=1}$, E0 is $E^{z=0}$, Y00 is $Y^{z=0,e=0}$, Y01 is $Y^{z=0,e=1}$, Y10 is $Y^{z=1,e=0}$, and Y11 is $Y^{z=1,e=1}$. The following three exercises are based on this population.

Ex 4.1

Use the treatment policy strategy to handle ICE E and find the value of the following estimand:

$$\theta^*_{\text{TP}} = \frac{1}{N}\sum_{i=1}^{N}\{Y_i^{z=1} - Y_i^{z=0}\}.$$

Ex 4.2

Use the hypothetical strategy to handle ICE E and find the value of the following estimand:

$$\theta^*_{\text{H}} = \frac{1}{N}\sum_{i=1}^{N}\left\{Y_i^{z=1,e=0} - Y_i^{z=0,e=0}\right\}.$$

Ex 4.3

Use the principal stratum strategy to handle ICE E. Find the value of the following estimand:

$$\theta^*_{\text{PS}_{00}} = \frac{1}{\#(\text{PS}_{00})}\sum_{i\in\text{PS}_{00}}\{Y_i^{z=1} - Y_i^{z=0}\},$$

where $\#(\text{PS}_{00})$ is the size of PS_{00}.

5

Longitudinal Studies

5.1 Continuous or Binary Outcome

In the first four chapters, we focused on studies where there is only one follow-up time and there is no time-dependent covariates, as illustrated in Figure 1.2. In this chapter, we will discuss longitudinal studies. In longitudinal studies with time-dependent treatments or intercurrent events (ICEs), we need to incorporate time explicitly in the definition of treatment (Hernán and Robins 2020).

Assume that there is one longitudinal study starting from baseline $t = 0$, along with follow-up visits, $t = 1, \ldots, T$, as illustrated in Figure 5.1. Assume that the primary outcome variable, Y, which is either continuous or binary, is the outcome variable measured at the final visit T.

At baseline $t = 0$, let Z be the treatment assignment, with $Z = 1$ and $Z = 0$ standing for assignment to treatment 1 and treatment 0, respectively, either by complete randomization or stratified randomization conditional on factor S. Let $X(0)$ be the vector containing baseline characteristics, including stratification factor S, if any.

To describe the time-dependent treatment explicitly, we introduce the notation $A(t)$, $t = 0, \ldots, T-1$. Between baseline $t = 0$ and follow-up visit $t = 1$, let $A(0)$ be the treatment that is actually taken by the patient. Some possible values that $A(t), t = 0, \ldots, T-1$, may take on are:

- 1: Treatment 1,
- 0: Treatment 0,
- `NULL`: No treatment,
- 1+: Treatment 1 with some add-on,
- 0+: Treatment 0 with some add-on,
- 2: Alternative treatment different from 1 and 0.

At follow-up visit $t = 1$, let $X(1)$ be the vector containing time-dependent covariates and/or the intermediate outcome variable measured at $t = 1$.

Similarly, let $A(t-1)$ be the treatment that is actually taken between visit $t-1$ and visit t, and let $X(t)$ be the vector containing time-dependent covariates and/or the intermediate outcome variable measured at visit t, for $t = 1, \ldots, T-1$.

DOI: 10.1201/9781003433378-5

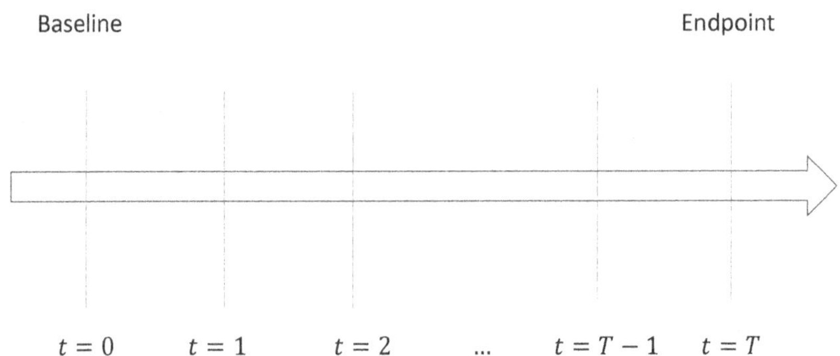

FIGURE 5.1
Baseline and follow-up visits

Finally, let $A(T-1)$ be the treatment that is actually taken between visit $T-1$ and visit T, and let Y be the primary outcome measured at visit T.

Let $\overline{A}(t) = (A(0), \ldots, A(t))$ be the actually taken treatment sequence up to t. Let $\overline{X}(t) = (X(0), \ldots, X(t))$ be the vector consisting of all the observed history—including baseline covariates, time-dependent covariates, and intermediate outcomes—up to time t, $t = 0, \ldots, T-1$. In particular, define $\overline{A} = (A(0), \ldots, A(T-1))$ and $\overline{X} = (X(0), \ldots, X(T-1))$.

5.1.1 The intent-to-treat effect

Causal estimand

Let $\Delta(t)$ be the indicator of analysis dropout at time t, $t = 1, \ldots, T$. For the purpose of demonstration, we assume that the analysis dropout pattern is monotone; that is, if $\Delta(t) = 1$, then $\Delta(t') = 1$ for $t' = t+1, \ldots, T$. The methods to be discussed in this chapter can be applied to both monotone patterns and non-monotone patterns. Let $\overline{\Delta} = (\Delta(1), \ldots, \Delta(T))$ and $\overline{\Delta}(t) = (\Delta(1), \ldots, \Delta(t))$, $t = 1, \ldots, T$.

Let $Y^{z=j,\overline{\delta}=\overline{0}}$ be the potential outcome had the patient been assigned to treatment j and the data been collected throughout the study period, $j = 1, 0$.

Similar to primary outcome Y, $X(t)$, $t = 1, \ldots, T-1$, are post-treatment initiation variables, and therefore we can define their potential outcomes. Let $X^{z=j,\overline{\delta}(t)=\overline{0}}(t)$ be the potential outcome had the patient been assigned to treatment j and the data been collected up to time t, $j = 1, 0$. Here, by convention, $\overline{0}$ is a vector of all zeros and its dimension is the same as the one of $\overline{\delta}(t)$. Thus, define the set of all potential outcomes as

$$W^{z=j} = \{X^{z=j,\overline{\delta}(t)=\overline{0}}(t), t = 1, \ldots, T-1; Y^{z=j,\overline{\delta}=\overline{0}}\}, j = 1, 0. \tag{5.1}$$

Following the discussion in Chapter 4, as illustrated in Figure 5.2, if we use the treatment policy strategy to handle treatment dropouts if any, along with

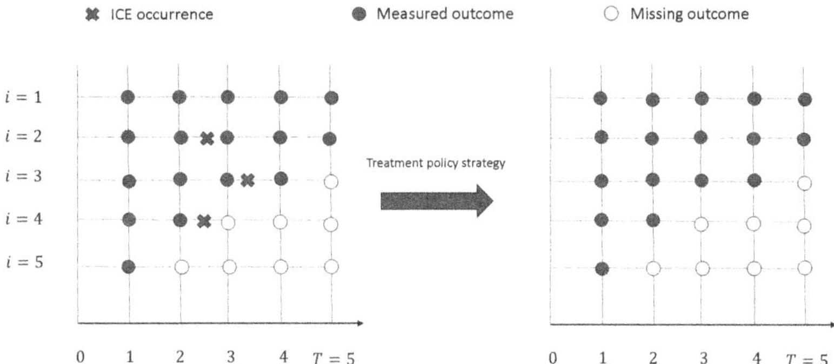

FIGURE 5.2
The treatment policy strategy to handle treatment dropouts

the hypothetical strategy to handle analysis dropouts (the source of missing data), we are interested in the intent-to-treat (ITT) estimand,

$$\theta^*_{\text{ITT}} = \mathbb{E}\left(Y^{z=1,\bar{\delta}=\bar{0}} - Y^{z=0,\bar{\delta}=\bar{0}}\right). \tag{5.2}$$

Statistical estimand

In order to translate the above causal estimand into a statistical estimand, we make the following three assumptions:

1. The consistency assumption:

$$\begin{aligned} X(t) &= X^{z=j,\bar{\delta}(t)=\bar{0}}(t), \text{ if } Z = j, \overline{\Delta}(t) = \bar{0}, \\ Y &= Y^{z=j,\bar{\delta}=\bar{0}}, \text{ if } Z = j, \overline{\Delta} = \bar{0}; \end{aligned}$$

2. The missing at random (MAR) assumption:

$$\begin{aligned} &(W^{z=1}, W^{z=0}) \perp\!\!\!\perp \Delta(1) | \left(Z, X(0)\right), \\ &(W^{z=1}, W^{z=0}) \perp\!\!\!\perp \Delta(t) | \left(Z, \overline{X}(t-1), \overline{\Delta}(t-1) = \bar{0}\right), t = 1, \dots, T-1; \end{aligned}$$

3. The positivity assumption:

$$\begin{aligned} &\mathbb{P}\left(\Delta(t) = 0 | Z = j, \overline{X}(t-1) = \overline{x}(t-1), \overline{\Delta}(t-1) = \bar{0}\right) > 0, \\ &\quad \text{for } t = 1, \dots, T-1; j = 1, 0; \overline{x}(t-1) \in \text{supp}(\overline{X}(t-1)). \end{aligned}$$

Under these three assumptions, we will show that

$$\mathbb{E}\left(Y^{z=j,\bar{\delta}=\bar{0}}\right) = \mathbb{E}_{\overline{X} \sim \mathcal{F}_j}\left[\mathbb{E}(Y | \overline{X}, Z = j, \overline{\Delta} = \bar{0})\right], \tag{5.3}$$

where the outer expectation on the right-hand-side (RHS) is over $\overline{X} \sim \mathcal{F}_j$ and $\mathcal{F}_j$ is a distribution function of $(X(0), \ldots, X(T-1))$ defined as

$$\begin{aligned}
&\mathbb{P}_{\mathcal{F}_j}(X(0) = x(0), \ldots, X(T-1) = x(T-1)) \\
=&\mathbb{P}\{X(0) = x(0)\} \prod_{t=1}^{T-1} \mathbb{P}\left\{X(t) = x(t) | \overline{X}(t-1) = \overline{x}(t-1), Z = j, \overline{\Delta}(t) = \overline{0}\right\}.
\end{aligned}$$

For a study where $T = 1$, (5.3) becomes

$$\mathbb{E}\left(Y^{z=j,\delta(1)=0}\right) = \mathbb{E}_{X(0)}\left[\mathbb{E}(Y|X(0), Z = j, \Delta(1) = 0)\right],$$

which has been shown in Chapter 3.

If (5.3) is proved, then we can complete the task of translating the causal estimand into a statistical estimand:

$$\begin{aligned}
\theta^*_{\text{ITT}} &= \mathbb{E}\left(Y^{z=j,\overline{\delta}=\overline{0}}\right) - \mathbb{E}\left(Y^{z=j,\overline{\delta}=\overline{0}}\right) \\
&= \mathbb{E}_{\overline{X}\sim\mathcal{F}_1}\left[\mathbb{E}(Y|\overline{X}, Z = 1, \overline{\Delta} = \overline{0})\right] - \mathbb{E}_{\overline{X}\sim\mathcal{F}_0}\left[\mathbb{E}(Y|\overline{X}, Z = 0, \overline{\Delta} = \overline{0})\right] \\
&\triangleq \theta_{\text{ITT}}. \qquad (5.4)
\end{aligned}$$

Before we provide formal proof for (5.3), let's understand the meaning of RHS of (5.3) using a numerical example where $T = 2$. Table 5.1 shows the observed data of a population of 20 subjects—imagine that one subject represents one million subjects.

Based on the data in Table 5.1, the complete-case subset consists of subjects 4–10, 12–13, 15, 17, and 10–20. Using the data from the complete-case subset, we obtain

$$\begin{aligned}
\mathbb{E}(Y|X(0) = 1, Z = 1, X(1) = 1, \overline{\Delta} = \overline{0}) &= 1/2, \\
\mathbb{E}(Y|X(0) = 1, Z = 1, X(1) = 0, \overline{\Delta} = \overline{0}) &= 1, \\
\mathbb{E}(Y|X(0) = 0, Z = 1, X(1) = 1, \overline{\Delta} = \overline{0}) &= 0, \\
\mathbb{E}(Y|X(0) = 0, Z = 1, X(1) = 0, \overline{\Delta} = \overline{0}) &= 1, \\
\mathbb{E}(Y|X(0) = 1, Z = 0, X(1) = 1, \overline{\Delta} = \overline{0}) &= 1/2, \\
\mathbb{E}(Y|X(0) = 1, Z = 0, X(1) = 0, \overline{\Delta} = \overline{0}) &= 0, \\
\mathbb{E}(Y|X(0) = 0, Z = 0, X(1) = 1, \overline{\Delta} = \overline{0}) &= 1/2, \\
\mathbb{E}(Y|X(0) = 0, Z = 0, X(1) = 0, \overline{\Delta} = \overline{0}) &= 0.
\end{aligned}$$

Based on all the observed data, we obtain distributions $\mathcal{F}_1$ and $\mathcal{F}_0$:

$$\begin{aligned}
&\mathbb{P}_{\mathcal{F}_1}(X(0) = 1, X(1) = 1) \\
&= \mathbb{P}(X(0) = 1)\mathbb{P}(X(1) = 1|X(0) = 1, Z = 1, \Delta(1) = 0) \\
&= (12/20)(3/4) = 9/20,
\end{aligned}$$

TABLE 5.1
Data from a population

SID	$X(0)$	Z	$X(1)$	Y
1	1	1	1	NA
2	1	1	NA	NA
3	1	1	NA	NA
4	1	1	0	1
5	1	1	1	0
6	1	1	1	1
7	1	0	0	0
8	1	0	0	0
9	1	0	1	1
10	1	0	1	0
11	1	0	NA	NA
12	1	0	0	0
13	0	1	1	0
14	0	1	1	NA
15	0	1	0	1
16	0	1	0	NA
17	0	0	1	0
18	0	0	NA	NA
19	0	0	1	1
20	0	0	0	0

$$
\begin{aligned}
&\mathbb{P}_{\mathcal{F}_1}(X(0)=1, X(1)=0) \\
&= \mathbb{P}(X(0)=1)\mathbb{P}(X(1)=0|X(0)=1, Z=1, \Delta(1)=0) \\
&= (12/20)(1/4) = 3/20, \\
&\mathbb{P}_{\mathcal{F}_1}(X(0)=0, X(1)=1) \\
&= \mathbb{P}(X(0)=0)\mathbb{P}(X(1)=1|X(0)=0, Z=1, \Delta(1)=0) \\
&= (8/20)(2/4) = 1/5, \\
&\mathbb{P}_{\mathcal{F}_1}(X(0)=0, X(1)=0) \\
&= \mathbb{P}(X(0)=0)\mathbb{P}(X(1)=0|X(0)=0, Z=1, \Delta(1)=0) \\
&= (8/20)(2/4) = 1/5;
\end{aligned}
$$

and

$$
\begin{aligned}
&\mathbb{P}_{\mathcal{F}_0}(X(0)=1, X(1)=1) \\
&= \mathbb{P}(X(0)=1)\mathbb{P}(X(1)=1|X(0)=1, Z=0, \Delta(1)=0) \\
&= (12/20)(2/5) = 6/25,
\end{aligned}
$$

$$
\begin{aligned}
&\mathbb{P}_{\mathcal{F}_0}(X(0)=1, X(1)=0)\\
&=\mathbb{P}(X(0)=1)\mathbb{P}(X(1)=0|X(0)=1, Z=0, \Delta(1)=0)\\
&=(12/20)(3/4)=9/25,\\
&\mathbb{P}_{\mathcal{F}_0}(X(0)=0, X(1)=1)\\
&=\mathbb{P}(X(0)=1)\mathbb{P}(X(1)=1|X(0)=0, Z=0, \Delta(1)=0)\\
&=(8/20)(2/3)=4/15,\\
&\mathbb{P}_{\mathcal{F}_0}(X(0)=0, X(1)=0)\\
&=\mathbb{P}(X(0)=1)\mathbb{P}(X(1)=0|X(0)=0, Z=0, \Delta(1)=0)\\
&=(8/20)(1/3)=2/15.
\end{aligned}
$$

Thus, we have

$$
\begin{aligned}
\mathbb{E}_{\overline{X}\sim\mathcal{F}_1}\left[\mathbb{E}(Y|\overline{X}, Z=1, \overline{\Delta}=\overline{0})\right] &= 9/20(1/2)+3/20(1)+1/5(0)+1/5(1)\\
&=23/40,\\
\mathbb{E}_{\overline{X}\sim\mathcal{F}_0}\left[\mathbb{E}(Y|\overline{X}, Z=0, \overline{\Delta}=\overline{0})\right] &= 6/25(1/2)+9/25(0)+4/15(1/2)+2/15(0)\\
&=19/75.
\end{aligned}
$$

Formula like (5.3) is usually referred to as the *g-computation algorithm*, also known as the *g-formula* (Hernán and Robins 2020; Robins 1986).

"The 'g' stands for 'generalized'."—Hernán and Robins (2020)

Now we are ready to prove g-formula (5.3). Although it is tedious, the proof is important for us to understand why we are able to translate causal estimand to statistical estimand in a longitudinal study. For simplicity, we consider $T=2$ and consider the setting where all the variables are discrete. The result can be easily extended to any $T\geq 2$ and the setting where there is a mixture of continuous variables with density functions and discrete variables with probability mass functions.

<u>*Proof of g-formula (5.3)*</u>: For $T=2$, we have

$$
\begin{aligned}
&\mathbb{P}\{Y=y|X(0)=x(0), Z=j, \Delta(1)=0, X(1)=x(1), \Delta(2)=0\}\\
&\overset{(a)}{=}\mathbb{P}\{Y^{z=j,\overline{\delta}(2)=\overline{0}}=y|\overline{X}(1)=\overline{x}(1), Z=j, \Delta(1)=0, \Delta(2)=0\}\\
&\overset{(b)}{=}\mathbb{P}\{Y^{z=j,\overline{\delta}(2)=\overline{0}}=y|X(0)=x(0), X(1)=x(1), Z=j, \Delta(1)=0\}\\
&\overset{(c)}{=}\mathbb{P}\{Y^{z=j,\overline{\delta}(2)=\overline{0}}=y|X(0)=x(0), X^{z=j,\delta(1)=0}(1)=x(1), Z=j, \Delta(1)=0\}\\
&\overset{(d)}{=}\frac{\mathbb{P}\{Y^{z=j,\overline{\delta}(2)=\overline{0}}=y, X^{z=j,\delta(1)=0}(1)=x(1)|X(0)=x(0), Z=j, \Delta(1)=0\}}{\mathbb{P}\{X^{z=j,\delta(1)=0}(1)=x(1)|X(0)=x(0), Z=j, \Delta(1)=0\}}\\
&\overset{(e)}{=}\frac{\mathbb{P}\{Y^{z=j,\overline{\delta}(2)=\overline{0}}=y, X^{z=j,\delta(1)=0}(1)=x(1)|X(0)=x(0), Z=j\}}{\mathbb{P}\{X^{z=j,\delta(1)=0}(1)=x(1)|X(0)=x(0), Z=j\}}\\
&\overset{(f)}{=}\frac{\mathbb{P}\{Y^{z=j,\overline{\delta}(2)=\overline{0}}=y, X^{z=j,\delta(1)=0}(1)=x(1)|X(0)=x(0)\}}{\mathbb{P}\{X^{z=j,\delta(1)=0}(1)=x(1)|X(0)=x(0)\}}\\
&\overset{(g)}{=}\mathbb{P}\{Y^{z=j,\overline{\delta}(2)=\overline{0}}=y|X(0)=x(0), X^{z=j,\delta(1)=0}(1)=x(1)\},
\end{aligned}
$$

where (a) holds under the consistency assumption, (b) holds under the MAR assumption, (c) holds under the consistency assumption again, (d) holds using the relationship among joint, marginal, and conditional distributions, (e) holds applying the MAR assumption in both the numerator and denominator, (f) holds because of the randomization, and (g) holds using the relationship among joint, marginal, and conditional distributions again.

We also have

$$
\begin{aligned}
&\mathbb{P}\{X(1) = x(1)|X(0) = x(0), Z = j, \Delta(1) = 0\}\\
\overset{(a)}{=}&\mathbb{P}\{X^{z=j,\delta(1)=0}(1) = x(1)|X(0) = x(0), Z = j, \Delta(1) = 0\}\\
\overset{(b)}{=}&\mathbb{P}\{X^{z=j,\delta(1)=0}(1) = x(1)|X(0) = x(0), Z = j\}\\
\overset{(c)}{=}&\mathbb{P}\{X^{z=j,\delta(1)=0}(1) = x(1)|X(0) = x(0)\},
\end{aligned}
$$

where (a) holds under the consistency assumption, (b) holds under the MAR assumption, and (c) holds because of the randomization.

Combining the above two results, we have

$$
\begin{aligned}
&\mathbb{P}\{X(0) = x(0), X^{z=j,\delta(1)=0}(1) = x(1), Y^{z=j,\bar{\delta}(2)=\bar{0}} = y\}\\
=&\mathbb{P}\{X(0) = x(0)\} \times \mathbb{P}\{X^{z=j,\delta(1)=0}(1) = x(1)|X(0) = x(0)\}\\
&\times \mathbb{P}\{Y^{z=j,\bar{\delta}(2)=\bar{0}} = y|X(0) = x(0), X^{z=j,\delta(1)=0}(1) = x(1)\}\\
=&\mathbb{P}\{X(0) = x(0)\} \times \mathbb{P}\{X(1) = x(1)|X(0) = x(0), Z = j, \Delta(1) = 0\}\\
&\times \mathbb{P}\{Y = y|X(0) = x(0), Z = j, \Delta(1) = 0, X(1) = x(1), \Delta(2) = 0\}.
\end{aligned}
$$

Writing the above finding in terms of expectation, we have

$$
\mathbb{E}\left(Y^{z=j,\bar{\delta}(2)=\bar{0}}\right) = \mathbb{E}_{\overline{X}(1)\sim\mathcal{F}_j}\left[\mathbb{E}(Y|\overline{X}(1), Z = j, \overline{\Delta}(2) = \bar{0})\right],
$$

where the outer expectation on the right-hand-side is over $\overline{X}(1) \sim \mathcal{F}_j$ and $\mathcal{F}_j$ is a distribution function of $(X(0), X(1))$ defined as

$$
\mathbb{P}\{X(0) = x(0)\} \times \mathbb{P}\{X(1) = x(1)|X(0) = x(0), Z = j, \Delta(1) = 0\}.
$$

This completes the proof for $T = 2$. □

Estimator

Define the following regression function,

$$
Q_{\overline{\Delta}=\bar{0}}(Z, \overline{X}) = \mathbb{E}(Y|\overline{X}, Z = 1, \overline{\Delta} = \bar{0}). \tag{5.5}
$$

Then the statistical estimand defined in (5.4) can be written as

$$
\theta_{\text{ITT}} = \mathbb{E}_{\overline{X}\sim\mathcal{F}_1}\left[Q_{\overline{\Delta}=\bar{0}}(1, \overline{X})\right] - \mathbb{E}_{\overline{X}\sim\mathcal{F}_0}\left[Q_{\overline{\Delta}=\bar{0}}(0, \overline{X})\right].
$$

Therefore, if we can obtain an estimator of $Q_{\overline{\Delta}=\overline{0}}(Z, \overline{X})$, denoted as $\widehat{Q}_{\overline{\Delta}=\overline{0}}(Z, \overline{X})$, and obtain the empirical distributions of $\mathcal{F}_1$ and $\mathcal{F}_0$, denoted as $\widehat{\mathcal{F}}_1$ and $\widehat{\mathcal{F}}_0$, respectively, then we can obtain an estimator for θ_{ITT},

$$\widehat{\theta}_{\text{ITT}} = \mathbb{E}_{\overline{X} \sim \widehat{\mathcal{F}}_1}\left[\widehat{Q}_{\overline{\Delta}=\overline{0}}(1, \overline{X})\right] - \mathbb{E}_{\overline{X} \sim \widehat{\mathcal{F}}_0}\left[\widehat{Q}_{\overline{\Delta}=\overline{0}}(0, \overline{X})\right]. \tag{5.6}$$

We relegate the discussion on whether there are better estimation methods and how to estimate their variances to Chapter 9. In this chapter, we focus on the task of identification from causal estimand to statistical estimand, along with a simple demonstration that there is at least one method—regardless of how good it is—to estimate the statistical estimand.

When T is large or $X(t)$ is high-dimensional, a more practical approach to constructing an estimator similar to (5.6) is multiple imputation (MI) (Rubin 1996, 2004). In practice, we can use statistical software (say, R package "mice" or SAS procedure "MI") to implement MI.

Behind each MI programming, there are implicit imputation models that fit $\mathcal{F}_1$ and $\mathcal{F}_0$ sequentially (that is, impute missing values in $X(t)$ based observed data up to t, $t = 1, \ldots, T-1$) and an implicit imputation model that fits the conditional distribution of Y given $\overline{X}$ and Z. Such imputation models can be specified through specifying the corresponding arguments.

We conclude this section with a brief description of the MI method. For $m = 1, \ldots, M$, using a certain MI programming, we impute the missing values for the following variables in order: $X(0), Z, X(1), \ldots, X(T-1), Y$. Then we have mth complete dataset, $\mathcal{O}(m) = \{(X_i(0), Z_i, Y_i^*), i = 1, \ldots, N\}$, where variables with an asterisk take on the observed values if observed and take on the imputed values if missing. Based on $\mathcal{O}(m)$, we construct an estimator, $\widehat{\theta}_{\text{ITT}}(m)$, along with an estimator of its variance, $\widehat{V}(m)$, using methods discussed in Chapter 2, depending on whether complete randomization or stratified randomization is implemented. Finally, the MI point estimator is

$$\widehat{\theta}_{\text{ITT, MI}} = \frac{1}{M} \sum_{m=1}^{M} \widehat{\theta}_{\text{ITT}}(m), \tag{5.7}$$

along with an estimator of its variance using Rubin's rule,

$$\widehat{V}_{\text{ITT, MI}} = \frac{1}{M} \sum_{m=1}^{M} \widehat{V}(m) + \left(1 + \frac{1}{M}\right) \frac{1}{M-1} \sum_{m=1}^{M} \left[\widehat{\theta}_{\text{ITT}}(m) - \widehat{\theta}_{\text{ITT, MI}}\right]^2.$$

5.1.2 The per-protocol effect

Causal estimand

In the previous subsection, we apply the treatment policy strategy to handle treatment dropout. In this subsection, we consider the hypothetical strategy to handle treatment dropout. For this aim, let $E = 1$ if there exists t such that $0 \leq t \leq T-1$ and $A(t) \neq Z$; let $E = 0$ if $A(t) = Z$ for any $0 \leq t \leq T-1$. In addition, define the first time when a treatment dropout occurs: $t_0(E) = \min\{t : A(t) \neq Z, 0 \leq t \leq T-1\}$ if $E = 1$; $t_0(E) = T$ if $E = 0$.

By the hypothetical strategy, we envisage a hypothetical scenario in which the treatment dropout would not occur. We define $\widetilde{\Delta}(t)$:

$$\widetilde{\Delta}(t) = 1 \text{ if } t > t_0(E) \text{ or } \Delta(t) = 1; \widetilde{\Delta}(t) = 0 \text{ otherwise.} \tag{5.8}$$

Let $\overline{\widetilde{\Delta}} = (\widetilde{\Delta}(1), \ldots, \widetilde{\Delta}(T))$ and $\overline{\widetilde{\Delta}}(t) = (\widetilde{\Delta}(1), \ldots, \widetilde{\Delta}(t))$, $t = 1, \ldots, T$. In other words, we consider the data that may be collected after treatment dropout as "missing" data, with quotation marks indicating that, although they may be observed, they are considered as irrelevant.

Let $Y^{z=j,\overline{\widetilde{\delta}}=\overline{0}}$ be the potential outcome had the patient been assigned to treatment j and followed the protocol throughout the study period, $j = 1, 0$.

Similar to primary outcome Y, $X(t)$, $t = 1, \ldots, T-1$, are post treatment initiation variables and therefore we can also define potential outcomes for them. Let $X^{z=j,\overline{\widetilde{\delta}}(t)=\overline{0}}(t)$ be the potential outcome had the patient been assigned to treatment j and followed the protocol up to time t, $j = 1, 0$. Thus, define the set of all potential outcomes as

$$\widetilde{W}^{z=j} = \{X^{z=j,\overline{\widetilde{\delta}}(t)=\overline{0}}(t), t = 1, \ldots, T-1; Y^{z=j,\overline{\widetilde{\delta}}=\overline{0}}\}, j = 1, 0. \tag{5.9}$$

Thus, as illustrated in Figure 5.3, if we apply the hypothetical strategy to handle both treatment dropouts (i.e., an ICE which is the source of "missing" data) and analysis dropouts (i.e., another ICE which is the source of missing data), we are interested in the following estimand,

$$\theta^*_{\text{PP}} = \mathbb{E}\left(Y^{z=1,\overline{\widetilde{\delta}}=\overline{0}} - Y^{z=0,\overline{\widetilde{\delta}}=\overline{0}}\right). \tag{5.10}$$

Statistical estimand

In order to translate the above causal estimand into a statistical estimand, we make the following three assumptions:

1. The consistency assumption:

$$X(t) = X^{z=j,\overline{\widetilde{\delta}}(t)=\overline{0}}(t), \text{ if } Z = j, \overline{\widetilde{\Delta}}(t) = \overline{0},$$
$$Y = Y^{z=j,\overline{\widetilde{\delta}}=\overline{0}}, \text{ if } Z = j, \overline{\widetilde{\Delta}} = \overline{0};$$

2. The "missing" at random (MAR) assumption:

$$(\widetilde{W}^{z=1}, \widetilde{W}^{z=0}) \perp\!\!\!\perp \widetilde{\Delta}(1) | (Z, X(0)),$$
$$(\widetilde{W}^{z=1}, \widetilde{W}^{z=0}) \perp\!\!\!\perp \widetilde{\Delta}(t) \Big| \left(Z, \overline{X}(t-1), \overline{\widetilde{\Delta}}(t-1) = \overline{0}\right), t = 1, \ldots, T-1;$$

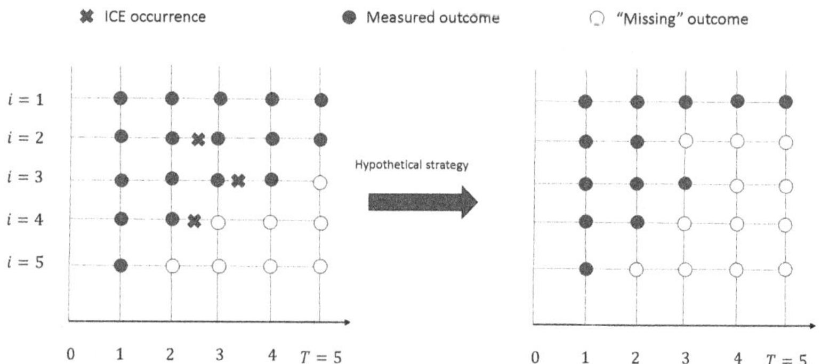

FIGURE 5.3
The hypothetical strategy to handle treatment dropouts

3. The positivity assumption:

$$\mathbb{P}\left(\widetilde{\Delta}(t)=0|Z=j,\overline{X}(t-1)=\overline{x}(t-1),\overline{\widetilde{\Delta}}(t-1)=\overline{0}\right)>0,$$

for any conditional event with non-zero probability.

Under these three assumptions, following the same arguments as those in the previous subsection with Δ replaced by $\widetilde{\Delta}$, we can show that

$$\mathbb{E}\left(Y^{z=j,\overline{\widetilde{\delta}}=\overline{0}}\right)=\mathbb{E}_{\overline{X}\sim\widetilde{\mathcal{F}}_j}\left[\mathbb{E}(Y|\overline{X},Z=j,\overline{\widetilde{\Delta}}=\overline{0})\right], \tag{5.11}$$

where $\widetilde{\mathcal{F}}_j$ is the following distribution function of $\overline{X}$,

$$\begin{aligned}&\mathbb{P}_{\widetilde{\mathcal{F}}_j}(X(0)=x(0),X(1)=x(1),\ldots,X(T-1)=x(T-1))\\=&\mathbb{P}\{X(0)=x(0)\}\prod_{t=1}^{T-1}\mathbb{P}\left\{X(t)=x(t)|\overline{X}(t-1)=\overline{x}(t-1),Z=j,\overline{\widetilde{\Delta}}(t-1)=\overline{0}\right\}.\end{aligned}$$

Thus, we show that, under the identifiability assumptions,

$$\begin{aligned}\theta^*_{\text{PP}}&=\mathbb{E}\left(Y^{z=1,\overline{\widetilde{\delta}}=\overline{0}}-Y^{z=0,\overline{\widetilde{\delta}}=\overline{0}}\right)\\&=\mathbb{E}_{\overline{X}\sim\widetilde{\mathcal{F}}_1}\left[\mathbb{E}(Y|\overline{X},Z=1,\overline{\widetilde{\Delta}}=\overline{0})\right]-\mathbb{E}_{\overline{X}\sim\widetilde{\mathcal{F}}_0}\left[\mathbb{E}(Y|\overline{X},Z=0,\overline{\widetilde{\Delta}}=\overline{0})\right]\\&\triangleq\theta_{\text{PP}}.\end{aligned} \tag{5.12}$$

Estimator

Following the same way of constructing estimator $\widehat{\theta}_{\text{ITT}}$ in (5.6) for θ_{ITT}, with Δ replaced by $\widetilde{\Delta}$, we can construct estimator $\widehat{\theta}_{\text{PP}}$ for θ_{PP}.

We also conclude this section with a brief description of the MI estimand. For $m = 1, \ldots, M$, using a certain MI programming, we impute the "missing" values for the following variables in order: $X(0), Z, X(1), \ldots, X(T-1), Y$. Then we have mth complete dataset, $\widetilde{\mathcal{O}}(m) = \{(X_i(0), Z_i, Y_i^*), i = 1, \ldots, N\}$, where variables with an asterisk are equal to the observed values if observed before any treatment dropout and take on imputed values if "missing." Based on $\widetilde{\mathcal{O}}(m)$, we construct an estimator $\widehat{\theta}_{\text{PP}}(m)$ along with an estimator of its variance, using methods discussed in Chapter 2, depending on whether complete randomization or stratified randomization is implemented. Finally, the MI point estimator is

$$\widehat{\theta}_{\text{PP, MI}} = \frac{1}{M} \sum_{m=1}^{M} \widehat{\theta}_{\text{PP}}(m), \tag{5.13}$$

along with an estimator of its variance Rubin's rule.

5.2 Time-to-event Outcome

Consider a longitudinal study with time-to-event outcomes. Motivated by Stitelman, De Gruttola, and Laan (2012) and Benkeser, Carone, and Gilbert (2018), we demonstrate that we can convert the time-to-event outcome to a longitudinally measured binary outcome.

Let Y be the time-to-event outcome of interest. Theoretically, Y can be considered continuous if there is no censoring, as discussed in Chapter 1. But in most studies with time-to-event outcomes, there is the problem of censoring, as discussed in Chapter 4. We demonstrate that we can convert the problem of censoring into the problem of analysis dropout.

As in the previous section, we consider a longitudinal study starting from baseline $t = 0$, followed by visits $t = 1, \ldots, T$.

At baseline $t = 0$, let Z be the treatment assignment, with $Z = 1$ and $Z = 0$ standing for assignment to treatment 1 and treatment 0, respectively, either by complete randomization or stratified randomization conditional on stratification factor S. Let $X(0)$ be the vector containing baseline characteristics, including stratification factor S, if any.

Between baseline $t = 0$ and follow-up visit $t = 1$, let $A(0)$ be the treatment that is actually taken by the patient.

At follow-up visit $t = 1$, let $Y(1)$ be the indicator of whether the primary event (PE) occurs at $t = 1$, with $Y(1) = 1$ standing for the occurrence of the PE and $Y(1) = 0$ otherwise. At the same time, let $\Delta(1)$ be the censoring status, with $\Delta(1) = 1$ and $\Delta = 0$ standing for being censored and not being censored, respectively. In addition, let $X(1)$ be the vector containing time-dependent covariates up to $t = 1$.

There is some difference between the missing data indicator discussed in the previous section and the censoring indicator discussed here. In the scenario of missing data, $\Delta(1) = 1$ means that the value of the outcome variable at $t = 1$ is missing. In the scenario of censoring, $\Delta(1) = 1$ means that the time-to-event outcome $Y > 1$; that is, although we don't know the exact value of Y, we know that the value of Y is larger than 1.

Similarly, for $t = 1, \ldots, T-1$, let $A(t-1)$ be the treatment that is actually taken between visit $t-1$ and visit t, let $Y(t)$ be the indicator of whether the PE occurs at visit t, let $\Delta(t)$ be the censoring status at visit t, and let $X(t)$ be the vector containing time-dependent covariates at visit t.

Finally, let $A(T-1)$ be the treatment that is actually taken between visit $T-1$ and final visit T, let $Y(T)$ be the indicator of whether the PE occurs at T, and let $\Delta(T)$ be the censoring status at T.

Again, there is some difference the between missing data indicator and the censoring indicator at the final visit T. In the scenario of missing data, $\Delta(T) = 1$ means that the value of the outcome variable at T is missing. In the scenario of censoring, the event that $\Delta(T) = 1$ and $\overline{\Delta}(T-1) = \overline{0}$ means that $Y > T$; that is, although we don't know the exact value of Y, we know that the value of Y is larger than T. For this reason, let $\overline{\Delta} = (\Delta(1), \ldots, \Delta(T-1))$.

There are two conventions. Convention one: If there is a t such that $1 \leq t \leq T$ and $Y(t) = 1$, then $Y(t') = 1$ and $\Delta(t') = 0$ for any $t' \geq t$. Convention two: if there is a t such that $1 \leq t \leq T$ and $\Delta(t) = 1$, then $\Delta(t') = 1$ and the value of $Y(t')$ is missing for any $t' > t$.

Figure 5.4 illustrates the discretization of time-to-event outcome Y into a longitudinally measured binary outcomes $Y(t)$, $t = 1, \ldots, T$.

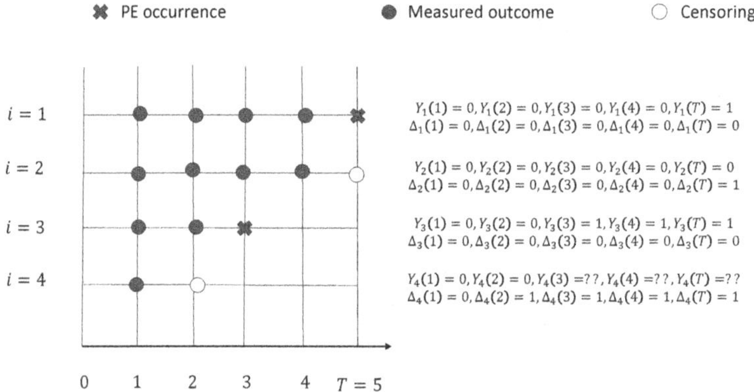

FIGURE 5.4
Discretization of time-to-event outcome

5.2.1 The intent-to-treat effect

Let $Y^{z=j,\bar{\delta}=\bar{0}}$ be the potential time to the PE had the patient been assigned to treatment j and the data been not censored up to visit $T-1$. Imagine that $Y^{z=j,\bar{\delta}=\bar{0}}$ would be discretized across $t = 1, \ldots, T$. Thus, let $Y^{z=j,\bar{\delta}=\bar{0}}(T)$ be the discretized potential outcome at the final visit T had the patient been assigned to treatment j and the data been not censored up to visit $T-1$, $j = 1, 0$. Define the following causal estimand:

$$\theta^*_{\text{ITT}} = \mathbb{P}\left[Y^{z=1,\bar{\delta}=\bar{0}} \leq T\right] - \mathbb{P}\left[Y^{z=0,\bar{\delta}=\bar{0}} \leq T\right], \tag{5.14}$$

which is the difference between the cumulative incidence function (CIF) at T had all the subjects been assigned to treatment 1 and not censored up to visit $T-1$ and the CIF at T had all the patients been assigned to treatment 0 and not censored up to visit $T-1$. It is equivalent to

$$\theta^*_{\text{ITT}} = \mathbb{P}\left\{Y^{z=1,\bar{\delta}=\bar{0}}(T) = 1\right\} - \mathbb{P}\left\{Y^{z=0,\bar{\delta}=\bar{1}}(T) = 1\right\}, \tag{5.15}$$

thanks to the discretization process. It is also equivalent to

$$\theta^*_{\text{ITT}} = \mathbb{P}\left\{Y^{z=0,\bar{\delta}=\bar{0}}(T) = 0\right\} - \mathbb{P}\left\{Y^{z=1,\bar{\delta}=\bar{0}}(T) = 0\right\}, \tag{5.16}$$

which is the difference between the survival rate at T had all the subjects been assigned to treatment 0 and not censored up to visit $T-1$ and the survival rate at T had all the patients been assigned to treatment 1 and not censored up to visit $T-1$.

From (5.15), we see that a study with time-to-event outcomes becomes a study with longitudinally measured binary outcomes. Thus, all the discussion on the estimand and estimator in the previous section still applies here.

To translate causal estimand (5.15) to a statistical estimand, we make the same three assumptions as those in Subsection 5.1.1, the consistency assumption, the MAR assumption—maybe better call it the censoring at random (CAR) assumption, and the positivity assumption. Under these assumptions, we can translate the causal estimand into a statistical estimand similar to (5.4) using the g-formula. Thus, we can apply the MI method to estimate the statistical estimand. We describe MI briefly in the following.

For $m = 1, \ldots, M$, using a certain MI programming (e.g., R package "mice" or SAS procedure "MI"), we impute the censored values for the following variables in order: $X(0), Z, X(1), Y(1), \ldots, X(T-1), Y(T-1), Y(T)$. Then we have mth complete dataset, $\mathcal{O}(m) = \{(X_i(0), Z_i, X_i^*(1), Y_i^*(1) \ldots, X_i^*(T-1), Y_i^*(T-1), Y_i^*(T)), i = 1, \ldots, N\}$, where variables with an asterisk are equal to the observed values if observed and take on imputed values if missing/censored. At this stage, there is some caveat: we need to adjust $\mathcal{O}(m)$ using the convention that if $Y^*(t) = 1$ for some t then set $Y^*(t') = 1$ for any $t' \geq t$. Denote the dataset after adjustment as $\mathcal{O}^*(m)$. Based on $\mathcal{O}^*(m)$, we obtain an estimator, $\widehat{\theta}_{\text{ITT}}(m)$, along with an estimator of its variance $\widehat{\widehat{V}}(m)$, using methods discussed in Chapter 2, depending on whether complete randomization or stratified randomization is implemented. For example,

$$\widehat{\theta}_{\text{ITT}}(m) = \frac{\sum_{i=1}^{N} Z_i Y_i^*(T)}{\sum_{i=1}^{N} Z_i} - \frac{\sum_{i=1}^{N}(1 - Z_i) Y_i^*(T)}{\sum_{i=1}^{N}(1 - Z_i)}. \tag{5.17}$$

Finally, we combine the M results using Rubin's rule.

In the above discussion, we consider survival rate or CIF as the population-level summary in defining the estimand. Alternatively, we may consider restricted mean survival time (RMST) as the population-level summary. Let $\theta^*_{\text{ITT, RMST}}$ be the estimand defined in terms of RMST limited by T. To estimate it, we revise the above MI method slightly. In the MI estimation, based on the dataset $\mathcal{O}^*(m)$, we obtain the following survival time limited by T,

$$T^*(m) = \min\{t : Y^*(t) = 1, 1 \le t \le T\} - 1,$$
$$T^*(m) = T, \text{ if set } \{t : Y^*(t) = 1, 1 \le t \le T\} \text{ is empty.}$$

Then we can construct the following MI point estimator for $\theta^*_{\text{ITT, RMST}}$,

$$\widehat{\theta}_{\text{ITT, RMST}}(m) = \frac{\sum_{i=1}^{N} Z_i T_i^*(m)}{\sum_{i=1}^{N} Z_i} - \frac{\sum_{i=1}^{N}(1 - Z_i) T_i^*(m)}{\sum_{i=1}^{N}(1 - Z_i)}. \tag{5.18}$$

5.2.2 The per-protocol effect

In the preceding subsection, we apply the treatment policy strategy to handle treatment dropout. In this subsection, we consider the hypothetical strategy to handle treatment dropout.

To apply the hypothetical strategy to handle treatment dropout, we only need to define $\widetilde{\Delta}(t)$ as in (5.8), which equals one if either treatment dropout or censoring occurs, whichever occurs first. In particular, let $\overline{\widetilde{\Delta}} = (\widetilde{\Delta}(1), \ldots, \widetilde{\Delta}(T-1))$. That is, we consider the data that may be collected after treatment dropout as "censored" data, with quotation marks indicating that they are observed but they are considered irrelevant.

Let $Y^{z=j,\overline{\widetilde{\delta}}=\overline{0}}(T)$ be the potential outcome had the patient been assigned to treatment j and following the protocol throughout the study period, $j = 1, 0$. Then, all the discussion of estimand and estimator is the same as that in the preceding subsection, except that $\Delta(t)$ is replaced by $\widetilde{\Delta}(t)$.

5.3 Treatment Regimes

The ITT effect is for the comparison between treatment assignment $Z = 1$ and treatment assignment $Z = 0$. The treatment policy strategy considers the

ICE as part of $Z = 1$ or $Z = 0$. Then, implicitly, $Z = 1$ and $Z = 0$ become two treatment policies. Thus, the ITT effect is for the comparison between treatment policy $Z = 1$ and treatment policy $Z = 0$. For example, if the anticipated ICE is treatment discontinuation, then treatment policy $Z = j$ means that the patient is assigned to treatment j and the patient may discontinue the treatment conditional on some event (e.g., lack of efficacy); that is, by the treatment policy strategy, the potential treatment discontinuation due to a lack of efficacy is part of the treatment policies under comparison, so treatment discontinuation is no longer considered as protocol deviation. As another example, if the anticipated ICE is switching to rescue medication, then treatment policy $Z = j$ means that the patient is assigned treatment to j and the patient may change the assigned treatment to rescue medication due to lack of efficacy; that is, by the treatment policy strategy, the potential switching to rescue medication due to lack of efficacy is part of the treatment policies under comparison, so switching to rescue medication is no longer considered as protocol deviation.

In using the treatment policy strategy, those ICEs (e.g., treatment discontinuation, switching to rescue medication) are not pre-specified in the definition of the treatment policies being compared; instead, they are implicitly incorporated as parts of the treatment policies being compared.

To make the treatment policies being compared more transparent, we can use the hypothetical strategy, by envisaging some hypothetical scenarios. So far, we have focused on only one hypothetical scenario, that is, the hypothetical scenario under which the ICE would not occur. Actually, we could envisage any number of hypothetical scenarios. For example, we could envisage the hypothetical scenario under which the patients would take no treatment after an ICE occurs. Note that one hypothetical scenario corresponds to a causal estimand; as long as the hypothetical scenario is pre-specified, the causal estimand is well-defined.

A better way to make the treatment policies being compared more transparent is by specifying treatment policies explicitly. Such treatment policies are called *dynamic treatment regimes* (DTRs)—including static treatment regimes (STRs) as special cases. For example, in Subsection 5.2.1, the two pre-specified STRs being compared are $\overline{a} = \overline{1}$ vs. $\overline{a} = \overline{0}$.

Now let's see how we can incorporate ICEs into the definition of DTRs. For example, to incorporate ICE "treatment discontinuation," we define two DTRs, "start with treatment 1 at time $t = 0$ and discontinue the treatment if lack of efficacy is shown at $t = 1$" and "start with treatment 0 at time $t = 0$ and discontinue the treatment if lack of efficacy is shown at $t = 1$." As another example, to incorporate ICE "rescue medication," we define two DTRs, "start with treatment 1 at time $t = 0$ and switch to rescue medication if lack of efficacy is shown at $t = 1$" and "start with treatment 0 at time $t = 0$ and switch to rescue medication if a lack of efficacy is shown at $t = 1$."

5.3.1 Dynamic treatment regimes

In this subsection, we only demonstrate how to define DTRs and how to define causal estimand in terms of potential outcomes. Interested readers may refer to the monograph entitled "Dynamic treatment regimes: statistical methods for precision medicine" by Tsiatis et al. (2020) for a systematic introduction to the methodology for DTRs.

Dynamic treatment regimes are within broader fields such as precision medicine and personalized medicine. As discussed by Tsiatis et al. (2020), one of the first uses of the term "Dynamic Treatment Regime" was by Murphy et al. (2001), who defined dynamic treatment regimes as comprising "rules for how the treatment level and type should vary with time." Tsiatis et al. (2020) developed R package "DynTxRegime" for implementing the methods introduced in their book.

For the purpose of demonstration, consider two-decision DTRs, where $T = 2$, with decision 0 to be made at $t = 0$, decision 1 to be made at $t = 1$, and the primary outcome to be observed at $t = 2$. The discussion can be easily extended to general DTRs with $T > 2$.

At baseline $t = 0$ when decision 0 is to be made, let $X(0)$ be information available and let $H(0)$ be the history accrued up to time $t = 0$; in this case, $H(0) = X(0)$. Let $A(0) \in \mathcal{A}(0)$ be the treatment option administered at decision 0. For example, $\mathcal{A}(0) = \{0, 1\}$ means that the treatment options at decision 0 are treatment 0 and treatment 1.

A decision rule at decision 0 is a function, $d_0(\cdot)$, that maps the patient's history to a treatment option in $\mathcal{A}(0)$:

$$d_0(\cdot) : \mathcal{H}(0) \to \mathcal{A}(0), \tag{5.19}$$

where $\mathcal{H}(0)$ is the support of $H(0)$.

For example, let $H(0) = X(0)$ be the disease severity, which may be "severe" or "moderate or mild"; that is, $\mathcal{H}(0) = \{$"severe," "moderate or mild"$\}$. We may consider decision rule, $d_0($"severe"$) = 1$ and $d_0($"moderate or mild"$) = 0$, which is to treat severe patients with treatment 1 and treat moderate or mild patients with treatment 0. We may consider static decision rules: $d_0(\cdot) \equiv 1$ is the decision rule to treat all patients with treatment 1, while $d_0(\cdot) \equiv 0$ is the decision rule to treat all patients with treatment 0.

At time $t = 1$ when decision 1 is to be made, let $X(1)$ be the additional information available between $t = 0$ and $t = 1$ and let $H(1)$ be the history accrued up to time $t = 1$; that is, $H(1) = \{X(0), A(0), X(1)\}$, where $A(0)$ is the treatment option that is actually taken by the patient. Let $A(1) \in \mathcal{A}(1)$ be the treatment option administered at decision 1. For example, $\mathcal{A}(1) = \{0, 1, 0+, 1+, \texttt{NULL}, 2\}$ means that the treatment options at decision 1 are treatment 0, treatment 1, treatment 0 with add-on, treatment 1 with add-on, no treatment, and alternative treatment 2.

A decision rule at decision 1 is a function, $d_1(\cdot)$, that maps the patient's history to a treatment option in $\mathcal{A}(1)$:

$$d_1(\cdot) : \mathcal{H}(1) \rightarrow \mathcal{A}(1), \tag{5.20}$$

where $\mathcal{H}(1)$ is the support of $H(1)$.

For example, let $H(2) = \{X(0), A(0), X(1)\}$, where $X(0)$ is the disease severity and $X(1)$ indicates whether the patient responses to the treatment $A(0)$. We may consider the following decision rule at $t = 1$,

$$d_1\{A(0) = 1, X(1) = \text{“response”}\} = \texttt{NULL}, \tag{5.21}$$

$$d_1\{A(0) = 1, X(1) = \text{“non-response”}\} = 1+, \tag{5.22}$$

$$d_1\{A(0) = 0, X(1) = \text{“response”}\} = \texttt{NULL}, \tag{5.23}$$

$$d_1\{A(0) = 0, X(1) = \text{“non-response”}\} = 0+. \tag{5.24}$$

This decision rule means that the patient receives treatment j at $t = 0$, followed by treatment j with add-on at $t = 1$ if non-response (i.e., the initial treatment is not effective), or by treatment discontinuation at $t = 1$ if response (i.e., the initial treatment is effective). Therefore, we are able to specify two treatment policies under comparison: treatment policy 1, $\overline{d}^{(1)} = (d_0^{(1)}(\cdot), d_1^{(1)}(\cdot))$ with $d_0^{(1)}(\cdot) \equiv 1$ and $d_1^{(1)}(\cdot) = d_1(\cdot)$ that is defined in (5.21)–(5.24), versus treatment policy 0, $\overline{d}^{(0)} = (d_0^{(0)}(\cdot), d_1^{(0)}(\cdot))$ with $d_0^{(0)}(\cdot) \equiv 0$ and $d_1^{(0)}(\cdot) = d_1(\cdot)$ that is defined in (5.21)–(5.24). In other words, treatment policy 1 means that the patient starts with treatment 1 at $t = 0$, followed by treatment 1 with add-on at $t = 1$ if non-response or followed by treatment discontinuation at $t = 1$ if response; while treatment policy 0 means that the patient starts with treatment 0 at $t = 0$, followed by treatment 0 with add-on at $t = 1$ if non-response or followed by treatment discontinuation at $t = 1$ if response.

We are ready to define the estimand of interest. Let $Y^{\overline{d}}$ be the potential outcome that would be achieved if the patient were to receive treatment sequence $\overline{d} = (d_0(\cdot), d_1(\cdot))$. For any two well-defined DTRs (in other words, any two well-defined treatment policies), denoted as $\overline{d}^{(1)}$ vs. $\overline{d}^{(0)}$, we may be interested in the following estimand:

$$\theta^*_{\text{DTR}} = \mathbb{E}\left\{Y^{\overline{d}^{(1)}}\right\} - \mathbb{E}\left\{Y^{\overline{d}^{(0)}}\right\}. \tag{5.25}$$

5.3.2 SMART design

We have seen that we can apply the treatment policy strategy to handle ICEs—reactively. Now we will see that we can apply the sequential multiple assignment randomized trial (SMART) to handle ICEs—proactively.

Figure 5.5 shows the schematic for a SMART with $T = 2$ stages designed to study the effects of several treatment sequences, motivated by several figures in Tsiatis et al. (2020). At stage 0, patients are randomized to receive

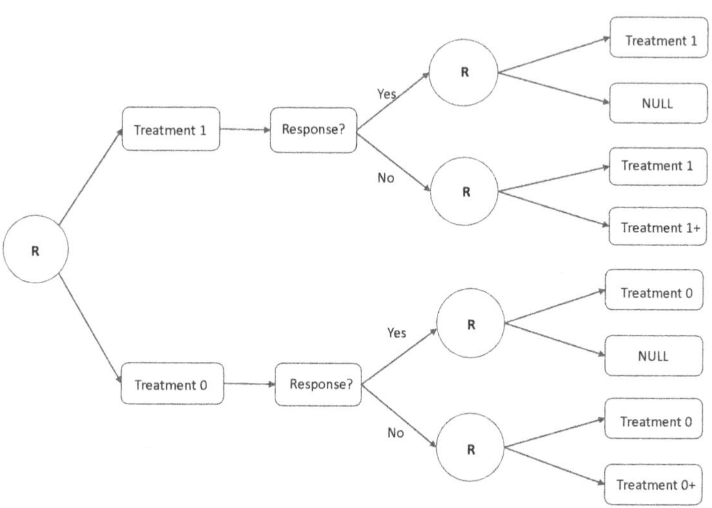

FIGURE 5.5
An example of SMART design where 'R' indicates randomization

treatment 1 or treatment 0. Whether or not a patient responds, which may be defined according to whether an outcome is improved by a certain percentage from baseline, is ascertained at the end of stage 0. At stage 1, the responders to treatment 1 are randomized to continuing treatment 1 or discontinuing treatment 1 (denoted as `NULL`), while non-responders to treatment 1 are randomized to continuing treatment 1 or starting treatment 1 with add-on (denoted as treatment 1+). Meanwhile, at stage 1, the responders to treatment 0 are randomized to continuing treatment 0 or discontinuing treatment 0 (denoted as `NULL`), while non-responders to treatment 0 are randomized to continuing treatment 0 or starting treatment 0 with add-on (denoted as treatment 0+).

In this SMART design, two stages of randomization are conducted sequentially, with the second stage of randomization depending on the outcome of the first stage of randomization.

Then, we are able to compare the following eight DTRs: (1) start with treatment 1 at $t = 0$ and continue treatment 1 at $t = 1$; (2) start with treatment 1 at $t = 0$, then continue treatment 1 if response or enhance with treatment 1+ if non-response; (3) start with treatment 1 at $t = 0$, then discontinue treatment 1 if response or continue treatment 1 if non-response; (4) start with treatment 1 at $t = 0$, then discontinue treatment 1 if response or enhance with treatment 1+ if non-response; (5) start with treatment 0 at $t = 0$ and continue treatment 0 at $t = 1$; (6) start with treatment 0 at $t = 0$, then continue treatment 0 if response or enhance with treatment 0+ if non-response; (7) start with treatment 0 at $t = 0$, then discontinue treatment 0 if response or continue treatment 0 if non-response; (8) start with treatment 0 at $t = 0$, then discontinue treatment 0 if response or enhance with treatment 0+ if non-response.

$$\begin{aligned}
\bar{d}^{(1)} &= \{d_0^{(1)} = 1, d_1^{(1)} = 1\}, \\
\bar{d}^{(2)} &= \{d_0^{(2)} = 1, d_1^{(2)}(\text{response}) = 1; d_1^{(2)}(\text{non-response}) = 1+\}, \\
\bar{d}^{(3)} &= \{d_0^{(3)} = 1, d_1^{(3)}(\text{response}) = \texttt{NULL}; d_1^{(3)}(\text{non-response}) = 1\}, \\
\bar{d}^{(4)} &= \{d_0^{(4)} = 1, d_1^{(4)}(\text{response}) = \texttt{NULL}; d_1^{(4)}(\text{non-response}) = 1+\}, \\
\bar{d}^{(5)} &= \{d_0^{(5)} = 0, d_1^{(5)} = 0\}, \\
\bar{d}^{(6)} &= \{d_0^{(6)} = 0, d_1^{(6)}(\text{response}) = 0; d_1^{(6)}(\text{non-response}) = 0+\}, \\
\bar{d}^{(7)} &= \{d_0^{(7)} = 0, d_1^{(7)}(\text{response}) = \texttt{NULL}; d_1^{(7)}(\text{non-response}) = 0\}, \\
\bar{d}^{(8)} &= \{d_0^{(8)} = 0, d_1^{(8)}(\text{response}) = \texttt{NULL}; d_1^{(8)}(\text{non-response}) = 0+\}.
\end{aligned}$$

After the above DRTs are defined, we are able to compare any two given DTRs, and furthermore, we are able to select the optimal DTR. Refer to Tsiatis et al. (2020) for a unified discussion.

5.4 Exercises

Ex 5.1

In Subsection 5.1.1, we use the standardization strategy to translate causal estimand θ^*_{ITT} into a statistical estimand θ_{ITT}. Alternatively, we can use the weighting strategy to do the job. We can show that, for $T = 2$,

$$
\begin{aligned}
&\mathbb{E}\left[\frac{I(Z=j,\Delta(1)=0,\Delta(2)=0)Y}{\mathbb{P}\{Z=j|X(0)\}\mathbb{P}\{\Delta(1)=0|X(0),Z=j\}\mathbb{P}\{\Delta(2)=0|X(0),Z=j,X(1)\}}\right]\\
&\overset{(a)}{=}\mathbb{E}\left[\frac{I(Z=j,\overline{\Delta}=0)Y}{\mathbb{P}\{Z=j|X(0)\}\pi_j\{X(0)\}\pi_j\{X(0),X(1)\}}\right]\\
&\overset{(b)}{=}\mathbb{E}\left[\frac{I(Z=j,\overline{\Delta}=0)Y^{z=j,\overline{\delta}=0}}{\mathbb{P}\{Z=j|X(0)\}\pi_j\{X(0)\}\pi_j\{X(0),X^{z=j,\delta(1)=0}(1)\}}\right]\\
&\overset{(c)}{=}\mathbb{E}\left\{\mathbb{E}\left[\frac{I(Z=j,\overline{\Delta}=0)Y^{z=j,\overline{\delta}=0}}{\mathbb{P}\{Z=j|X(0)\}\pi_j\{X(0)\}\pi_j\{X(0),X^{z=j,\delta(1)=0}(1)\}}\middle| X(0),W^{z=j}\right]\right\}\\
&\overset{(d)}{=}\mathbb{E}\left\{\frac{\mathbb{P}\{Z=j,\overline{\Delta}=0|X(0),X^{z=j,\delta(1)=0}(1)\}Y^{z=j,\overline{\delta}=0}}{\mathbb{P}\{Z=j|X(0)\}\pi_j\{X(0)\}\pi_j\{X(0),X^{z=j,\delta(1)=0}(1)\}}\right\}\\
&\overset{(e)}{=}\mathbb{E}\left\{Y^{z=j,\overline{\delta}=0}\right\}.
\end{aligned}
$$

Write down the reasons why equalities (a)–(e) hold.

Ex 5.2

In Subsection 5.1.1, based on the population shown in Table 5.1, we use the standardization strategy to calculate the value of statistical estimand θ_{ITT}. Based on the same population, use the weighting strategy proposed in Ex 5.1 to calculate the value of the following statistical estimand θ_{ITT},

$$
\mathbb{E}\left[\frac{I(Z=1,\overline{\Delta}=0)Y}{\mathbb{P}\{Z=1|X(0)\}\pi_j\{X(0)\}\pi_j\{\overline{X}\}}\right]-\mathbb{E}\left[\frac{I(Z=0,\overline{\Delta}=0)Y}{\mathbb{P}\{Z=0|X(0)\}\pi_j\{X(0)\}\pi_j\{\overline{X}\}}\right].
$$

Ex 5.3

In Section 5.2, we propose a method to discretize a time-to-event outcome into a longitudinal binary outcome. Let Y be the time-to-death in months, C be the censoring time in months, $Y^* = \min(Y, C)$ be the observed variable, and Δ be the censoring indicator. Use this method to discretize five time-to-event observations, with (Y_i^*, Δ_i), $i = 1, \ldots, 5$, being $(5,1), (7,0), (3,0), (4,0), (4,1)$, into longitudinal binary observations, $(Y_i(1), Y_i(2), \ldots, Y_i(7))$, $i = 1, \ldots, 5$, respectively. Use `NA` to denote missing data.

6

Real-World Evidence Studies

6.1 RWE Studies

Randomized controlled clinical trials (RCTs) are the gold standard for evaluating the safety and efficacy of pharmaceutical drugs, as discussed in the first five chapters. However, in many cases their costs, duration, limited generalizability, and ethical or technical feasibility have caused some to look for real-world evidence (RWE) studies as alternatives.

In 2018, the Food and Drug Administration (FDA) released *Framework for FDA's Real-World Evidence Program* and provided definitions for real-world data (RWD) and RWE:

> "Real-World Data (RWD) are data relating to patient health status and/or the delivery of health care routinely collected from a variety of sources. Real-World Evidence (RWE) is the clinical evidence about the usage and potential benefits or risks of a medical product derived from analysis of RWD."—Framework for FDA's RWE Program

The framework for FDA's RWE program covered many types of studies that generate RWE, which are referred to as RWE studies in this book. In the following three subsections, we describe three main types of RWE studies.

6.1.1 Pragmatic RCTs

As in the framework for FDA's RWE program, we refer to the randomized controlled clinical trials (RCTs) discussed in the previous five chapters as traditional RCTs and refer to the other more flexible RCTs conducted in the real-world setting as pragmatic RCTs. As pointed out by Thorpe et al. (2009), traditional RCTs seek to answer the question, "*Can this intervention work under ideal conditions?*", whereas pragmatic RCTs seek to answer the question, "*Does this intervention work under usual conditions?*".

Thorpe et al. (2009) summarized the differences between pragmatic RCTs and traditional RCTs in 10 domains. For simplicity, here we compare them following the PROTECT checklist.

DOI: 10.1201/9781003433378-6

Population: In traditional RCTs, the population is defined via a set of inclusion/exclusion criteria, whereas in pragmatic RCTs, the population is often defined loosely, e.g., defined as a population consisting of all the patients who have certain of conditions. *Response/Outcome*: In traditional RCTs, the response/outcome variables may need specialized training or testing, whereas in pragmatic RCTs, the response/outcome variables can be assessed under the usual conditions. *Treatment/Exposure*: The treatment/exposure variable—a binary variable indicating either the investigative intervention or the comparator—is defined with more restrictions in traditional RCTs than in pragmatic RCTs. Often the comparator is placebo in traditional RCTs to ensure the trial is double-blinded, whereas the pragmatic RCTs are open-label, in which the comparator is often "standard of care" or "usual practice." Thus, in pragmatic RCTs, there are bigger proportions of intercurrent events (ICEs) such as treatment dropout and treatment switching than in traditional RCTs. *Counterfactual thinking*: Since there are more missing data and ICEs in pragmatic RCTs than in traditional RCTs, counterfactual thinking and causal thinking play a more important role. *Time*: In traditional RCTs, patients are followed with more frequent visits and more extensive data are collected than in pragmatic RCTs conducted in routine practice.

Despite these differences between them, the causal inference methods that have been developed for traditional RCTs are applicable to pragmatic RCTs. Moreover, the causal inference methods are highly demanded for pragmatic RCTs, given that pragmatic RCTs are conducted in routine practice and there would be more missing data and ICEs.

6.1.2 Observational studies

In RCTs (traditional RCTs or pragmatic RCTs), the treatments are assigned by the investigators and the treatment assignment function is known to the investigators. Let Z be the treatment assignment variable, with $Z = j$ standing for assignment to treatment j, $j = 1, 0$. For example, in a 1:1 RCT, $\mathbb{P}(Z = 1) = 0.5$.

In observational studies, the treatments are determined by the patients and their physicians. Let A be the treatment received by the patient and X be the pre-treatment variables, based on which the treatment decision is made. The treatment assignment function, $g(1|X) = \mathbb{P}(A = 1|X)$, is referred to as the *propensity score function* (Rosenbaum and Rubin 1983). The major difference between RCTs and observational studies is that in RCTs the propensity score function is known whereas in observational studies it is unknown. In observational studies, the investigators only know that A may depend on X, without knowing which variables are in X and how A depend on X.

Observational studies include but are not limited to cohort studies, cross-sectional studies, and case-control studies. In this chapter, we focus on cohort studies. As discussed by Fang et al. (2020), there are two scenarios of cohort studies: (1) prospective cohort studies that are designed to generate RWD;

and (2) retrospective cohort studies that are designed based on preexisting RWD sources (e.g., electronic health records, claims data, or registries).

6.1.3 Externally controlled trials

Externally controlled trials (ECTs) are single-arm clinical trials with an external control group. To understand external controls, we need to understand concurrent controls first. ICH E10 "Choice of control group in clinical trials" provided the following definition.

> "A concurrent control group is one chosen from the same population as the test group and treated in a defined way as part of the same trial that studies the test treatment, and over the same period of time. The test and control groups should be similar with regard to all baseline and on-treatment variables that could influence outcome, except for the study treatment."—ICH E10

Consequently, ICH E10 provided the definition of ECT.

> "An externally controlled trial is one in which the control group consists of patients who are not part of the same randomized study as the group receiving the investigational agent; i.e., there is no concurrently randomized control group. The control group is thus not derived from exactly the same population as the treated population."—ICH E10

There are two major sources of external controls. We may choose external controls from the placebo arms of some completed RCTs and refer to them as historical controls. We may choose external controls from real-world data and referred as external RWD controls.

In the remaining of this chapter, we focus on observational studies and ECTs. In the first five chapters, we used directed acyclic graphs (DAGs) occasionally to display the relationship among variables. For observational studies, we will rely more on DAGs. Revisit Subsection 1.4.2 for a brief introduction to DAGs or refer to Pearl (2009) and Hernán and Robins (2020) for a comprehensive discussion on DAGs.

6.2 Confounding Bias

We often hear people say something like: "There is bias due to non-randomization." Is this true? What is bias?

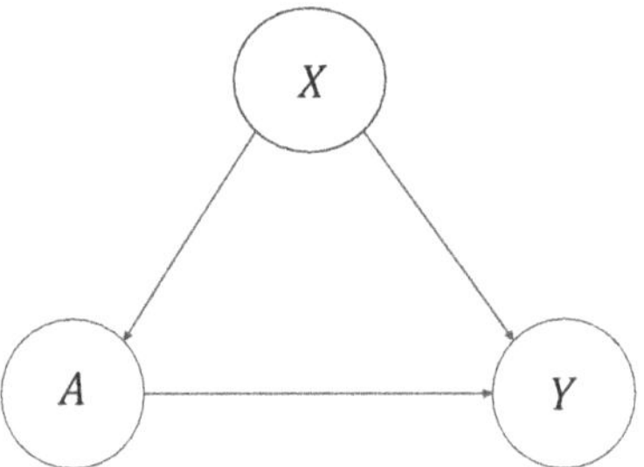

FIGURE 6.1
A well-conducted cohort study

We turn to ICH E9 "Statistical principles for clinical trials", which provided the following definition of bias:

> "As used in this guidance, the term 'bias' describes the systematic tendency of any factors associated with the design, conduct, analysis and interpretation of the results of clinical trials to make the estimate of a treatment effect deviate from its true value."—ICH E9

In the following three subsections, in three steps, we discuss the source of "confounding bias", using the example of a cohort study.

6.2.1 No unmeasured confounder

Assume there is one well-designed and well-conducted cohort study as displayed in Figure 6.1, assuming all the variables are measured without measurement bias, there are no missing data, and the sample size is large enough. In Figure 6.1, X is a vector of pre-treatment covariates, A is the treatment variable (taking on 1 or 0), and Y is the outcome variable (either continuous or binary).

Now let's see whether there is any confounding bias if we do causal inference in a certain way. When we speak of bias, we should first define the estimand of interest. Without referring to an estimand, any statement about bias is meaningless.

Causal estimand

Let $Y^{a=j}$ be the potential outcome that one patient would have had if the patient had been treated by treatment $j = 1$ or 0.

Recall that in an RCT, we start with an intent-to-treat (ITT) population of size N, which consists of data $(Y_i^{a=1}, Y_i^{a=0})$ in the potential world, $i = 1, \ldots, N$, which are assumed to be independent and identically distributed (i.i.d.) drawn from a super-population. That is, we say $(Y_i^{a=1}, Y_i^{a=0})$, $i = 1, \ldots, N$, are i.i.d. with $(Y^{a=1}, Y^{a=0})$.

Also, recall the population-sample duality discussed in Chapter 1. Therefore, in an observational study, we start with a **sample** of size n, which consists of data $(Y_i^{a=1}, Y_i^{a=0})$ in the potential world, $i = 1, \ldots, n$, which are assumed to be i.i.d. drawn from a **population**. We could construct such a population conceptually following the same way by which we construct the super-population in an RCT. Thus, we say $(Y_i^{a=1}, Y_i^{a=0})$, $i = 1, \ldots, n$, are i.i.d. with $(Y^{a=1}, Y^{a=0})$.

After we understand the above relationship between sample and population in an observational study, we are able to define an estimand of interest. For example, we are interested in the following causal estimand,

$$\theta^*_{\text{ATE}} = \mathbb{E}(Y^{a=1}) - \mathbb{E}(Y^{a=0}), \tag{6.1}$$

where the expectation is over the population. As suggested by the subscript, this causal estimand is referred to as the average treatment effect (ATE), which measures the difference between the mean of the potential outcomes if all the patients in the population had been treated by treatment 1 and that if all the patients in the population had been treated by treatment 0.

Identification

In the potential world, potential outcomes $(Y_i^{a=1}, Y_i^{a=0})$, $i = 1, \ldots, n$, are i.i.d. with $(Y^{a=1}, Y^{a=0})$. In the real world, observations (X_i, A_i, Y_i), $i = 1, \ldots, n$, are i.i.d. with (X, A, Y).

We make the following three identifiability assumptions.

1. The consistency assumption:

$$Y = Y^{a=j} \text{ if } A = j, \text{ for } j = 1, 0; \tag{6.2}$$

2. The exchangeability assumption:

$$(Y^{a=0}, Y^{a=1}) \perp\!\!\!\perp A | X; \tag{6.3}$$

3. The positivity assumption:

$$\mathbb{P}(A = j | X = x) > 0, \text{ for } j = 1, 0; x \in \text{supp}(X). \tag{6.4}$$

We are familiar with the consistency assumption and the positivity assumption, since we have seen them several times in some of the previous chapters. The exchangeability assumption is also known as the no-unmeasured-confounding (NUC) assumption (Hernán and Robins 2020). With the help of the structural causal model (SCM) in Appendix 6.5, we show that the exchangeability assumption is implied by the DAG in Figure 6.1.

Under these assumptions, we can translate the causal estiamnd into a statistical estimand. There are two strategies for this task: the standardization strategy and the weighting strategy (Fang 2020; Hernán and Robins 2020).

The standardization strategy

Under the identifiability assumptions, we can show that

$$\begin{aligned}\theta^*_{\text{ATE}} &= \mathbb{E}(Y^{a=1}) - \mathbb{E}(Y^{a=0})\\ &\overset{(a)}{=} \mathbb{E}[\mathbb{E}(Y^{a=1}|X)] - \mathbb{E}[\mathbb{E}(Y^{a=0}|X)]\\ &\overset{(b)}{=} \mathbb{E}[\mathbb{E}(Y^{a=1}|A=1,X)] - \mathbb{E}[\mathbb{E}(Y^{a=0}|A=0,X)]\\ &\overset{(c)}{=} \mathbb{E}[\mathbb{E}(Y|A=1,X)] - \mathbb{E}[\mathbb{E}(Y|A=0,X)],\end{aligned}$$

where (a) holds using the law of iterated expectation, (b) holds under the exchangeability assumption and the positivity assumption, and (c) holds under the consistency assumption. Thus, we translate the causal estimand to the following statistical estimand under the identifiability assumptions,

$$\theta_{\text{ATE}} = \mathbb{E}[\mathbb{E}(Y|A=1,X)] - \mathbb{E}[\mathbb{E}(Y|A=0,X)]. \tag{6.5}$$

As in Hernán and Robins (2020), we call the above identification method *the standardization strategy*. The other method is *the weighting strategy*.

The weighting strategy

Under the identifiability assumptions, we can show that

$$\begin{aligned}\mathbb{E}\left\{\frac{I(A=a)}{P(A=a|X)}Y\right\} &\overset{(a)}{=} \mathbb{E}\left\{\frac{I(A=a)}{\mathbb{P}(A=a|X)}Y^a\right\}\\ &\overset{(b)}{=} \mathbb{E}\left[\mathbb{E}\left\{\frac{I(A=a)}{\mathbb{P}(A=a|X)}Y^a\middle|X\right\}\right]\\ &\overset{(c)}{=} \mathbb{E}\left[\mathbb{E}\left\{\frac{I(A=a)}{\mathbb{P}(A=a|X)}\middle|X\right\}\mathbb{E}\{Y^a|X\}\right]\\ &\overset{(d)}{=} \mathbb{E}\left[\mathbb{E}\{Y^a|X\}\right]\\ &\overset{(e)}{=} \mathbb{E}(Y^a),\end{aligned}$$

where the left-hand-side of (a) is well-defined under the positivity assumption, the right-hand-side of (a) is using the consistency assumption, (b) holds using the law of iterated expectations, (c) holds under the exchangeability assumption, (d) holds because $\mathbb{E}\{I(A=a|X)\} = \mathbb{P}(A=a|X)$, and (e) holds using the law of iterated expectations Thus, under the identifiability assumptions,

$$\theta^*_{\text{ATE}} = \mathbb{E}\left[\frac{I(A=1)}{P(A=1|X)}Y\right] - \mathbb{E}\left[\frac{I(A=0)}{P(A=0|X)}Y\right].$$

Thus, by the weighting strategy, we translate the causal estimand to the following statistical estimand,

$$\theta_{\text{ATE}} = \mathbb{E}\left[\frac{I(A=1)}{P(A=1|X)}Y\right] - \mathbb{E}\left[\frac{I(A=0)}{P(A=0|X)}Y\right], \tag{6.6}$$

which is equivalent to the statistical etsimand defined in (6.5) under the same three identifiability assumptions.

Confounding bias

After we define a causal estimand and accomplish the task of identification, we are ready to discuss the issue of confounding bias.

The causal estimand, θ^*_{ATE}, is our ultimate target, while the statistical estimand, θ_{ATE}, is our realistic target.

In the next three chapters, we will discuss methods for constructing consistent estimators of θ_{ATE}. We denote such estimator as $\widehat{\theta}_{\text{ATE},n}$. We say that $\widehat{\theta}_{\text{ATE},n}$ is a consistent estimator of θ_{ATE} if

$$\widehat{\theta}_{\text{ATE},n} \xrightarrow{p} \theta_{\text{ATE}}, \text{ as } n \to \infty.$$

Since we have shown that $\theta^*_{\text{ATE}} = \theta_{\text{ATE}}$ under the identifiability assumptions, we conclude that if $\widehat{\theta}_{\text{ATE},n}$ is a consistent estimator of θ_{ATE} then it is also a consistent estimator of θ^*_{ATE}. Thus, if we use consistent estimator $\widehat{\theta}_{\text{ATE},n}$ to estimate θ^*_{ATE}, we say that there is **no confounding bias** under the identifiability assumptions.

However, what happens if we use the following naive estimator,

$$\widehat{\theta}'_n = \frac{\sum_{i=1}^n I(A_i=1)Y_i}{\sum_{i=1}^n I(A_i=1)} - \frac{\sum_{i=1}^n I(A_i=0)Y_i}{\sum_{i=1}^n I(A_i=0)},$$

to estimate θ^*_{ATE}? In this case, we say that there is statistical bias if we use $\widehat{\theta}'_n$ to estimate statistical estimand θ_{ATE}, because this naive estimator is a consistent estimator of the following estimand,

$$\theta' = \mathbb{E}(Y|A=1) - \mathbb{E}(Y|A=0),$$

which is not equal to θ_{ATE}. Furthermore, we say that there is **confounding bias** if we use $\widehat{\theta}'_n$ to estimate causal estimand θ^*_{ATE}, because θ' is not equal to θ^*_{ATE} even under the identifiability assumptions.

6.2.2 Unmeasured confounders

As displayed in Figure 6.2, assume that there is a vector of unmeasured confounders, denoted as U, along with measured variables X, A, and Y.

Let's see whether there is any confounding bias if we do causal inference in a certain way. Again, when we speak of bias, we should first define the estimand

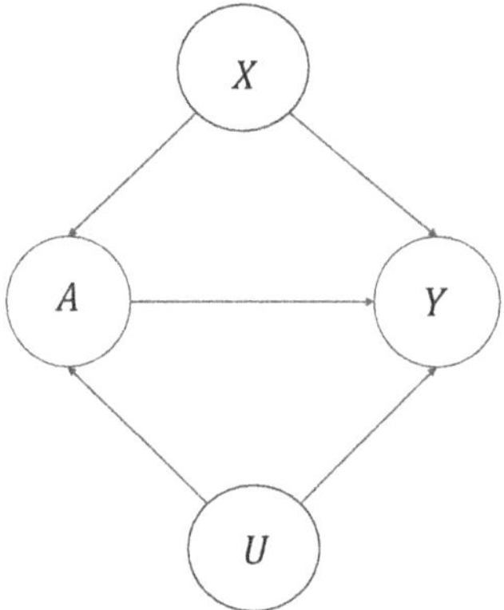

FIGURE 6.2
A cohort study with unmeasured confounders

of interest. Assume that we are interested in the same causal estimand, θ^*_{ATE}, defined in (6.1)

Next, for the purpose of identification, we reevaluate those three identifiability assumptions in the previous subsection. The consistency assumption and the positivity assumption are still applicable, but it is a problem to assume the exchangeability assumption because now we have

$$(Y^{a=0}, Y^{a=1}) \perp\!\!\!\perp A_i | (X, U). \tag{6.7}$$

Under the consistency assumption, the positivity assumption, and (6.7), we can show that

$$\begin{aligned}
\theta^*_{\text{ATE}} &= \mathbb{E}(Y^{a=1}) - \mathbb{E}(Y^{a=0}) \\
&\overset{(a)}{=} \mathbb{E}[\mathbb{E}(Y^{a=1}|X, U)] - \mathbb{E}[\mathbb{E}(Y^{a=0}|X, U)] \\
&\overset{(b)}{=} \mathbb{E}[\mathbb{E}(Y^{a=1}|A = 1, X, U)] - \mathbb{E}[\mathbb{E}(Y^{a=0}|A = 0, X, U)] \\
&\overset{(c)}{=} \mathbb{E}[\mathbb{E}(Y|A = 1, X, U)] - \mathbb{E}[\mathbb{E}(Y|A = 0, X, U)],
\end{aligned}$$

where (a) holds using the law of iterated expectations, (b) holds under (6.7), and (c) holds under the consistency assumption. However, the rightmost term of the above equation,

$$\mathbb{E}[\mathbb{E}(Y|A = 1, X, U)] - \mathbb{E}[\mathbb{E}(Y|A = 0, X, U)],$$

still involves unmeasured U. Thus, we are unable to translate the causal estimand into an estimable statistical estimand that depends only on the distribution of measured variables (X, A, Y).

Therefore, we are encountered with the issue of confounding bias. The causal estimand, θ^*_{ATE}, is our ultimate target. Assume that estimator $\widehat{\theta}_{\text{ATE}}$ is consistent in estimating θ_{ATE}. On the one hand, if θ_{ATE} is our realistic target, then there is no statistical bias if we use $\widehat{\theta}_{\text{ATE}}$ to estimate it. On the other hand, if θ^*_{ATE} is the ultimate target, then there is **confounding bias** if we use $\widehat{\theta}_{\text{ATE}}$ to estimate it, because

$$\begin{aligned}\theta^*_{\text{ATE}} &= \mathbb{E}[\mathbb{E}(Y|A=1,X,U)] - \mathbb{E}[\mathbb{E}(Y|A=0,X,U)] \\ &\neq \mathbb{E}[\mathbb{E}(Y|A=1,X)] - \mathbb{E}[\mathbb{E}(Y|A=0,X)] = \theta_{\text{ATE}}.\end{aligned}$$

Since θ^*_{ATE} is not identifiable if the DAG in Figure 6.2 holds true, we are unable to estimate θ^*_{ATE} directly. A common practice is to obtain the main result of estimating θ_{ATE} under the exchangeability assumption, and then conduct sensitivity analysis to evaluate the robustness of the main result if the exchangeability assumption is violated (e.g., if there are unmeasured confounders). We will discuss sensitivity analysis in Chapter 10.

6.2.3 Proxy variables

We have completed the first two steps of causal thinking. In Step 1, we assume that there is no unmeasured confounder, so we are able to construct consistent estimators to estimate the causal estimand. How to construct consistent estimators with good properties is the main topic of Chapters 7-9. In Step 2, we assume that there are unmeasured confounders, so we are unable to construct consistent estimators to estimate the causal estimand. Instead, we construct consistent estimators to estimate the statistical estimand and then conduct sensitivity analysis to evaluate the impact of the confounding bias due to unmeasured confounders.

Now we are ready to discuss the third step of causal thinking. In Step 3, we consider a cohort study where there are unmeasured confounders U, along with measured proxy variables V and W, as displayed in Figure 6.3. In Figure 6.3, V is a *treatment-inducing proxy* and W is an *outcome-inducing proxy* (Tchetgen Tchetgen et al. 2020). With such proxy variables, Miao, Geng, and Tchetgen Tchetgen (2018) showed that we are able to accomplish the task of identification, given some rank assumptions. The existence of rich RWD in RWE studies makes it possible to search for and identify proxy variables (Kummerfeld, Lim, and Shi 2022).

The improvement from the DAG in Figure 6.2 to that in Figure 6.3 is significant: it turns a non-identifiable research question into an identifiable one, making the causal estimand estimable. There are two ways to achieve this. One way is to divide vector X into three sub-vectors, $\{V, X, W\}$, such that V and W satisfy the conditions implied by the DAG in Figure 6.3; that is, there is a path from X to V and a path from V to A, while there is a path from X to W and a path from W to Y. The other way is to enrich the vector X, by adding two other vectors, V and W, satisfying the conditions of being treatment-inducing proxy and outcome-inducing proxy, respectively.

Assume that we are interested in estimating causal estimand θ^*_{ATE} defined in (6.1). To demonstrate that θ^*_{ATE} is identifiable, Tchetgen Tchetgen

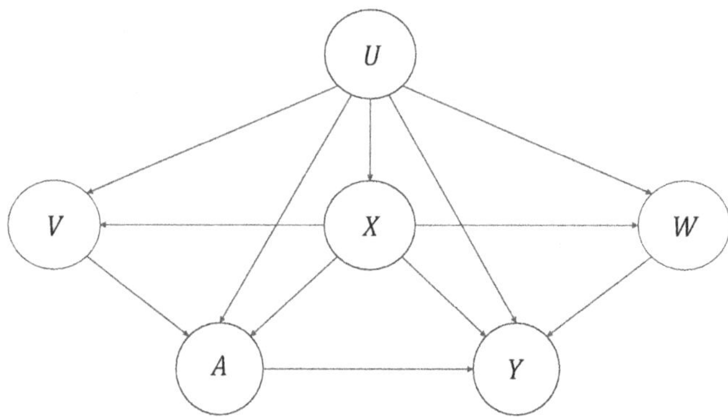

FIGURE 6.3
A cohort study with proxy variables

et al. (2020) considered structural linear models, respectively, for outcome variable Y and outcome-inducing proxy W:

$$\mathbb{E}(Y|A,X,V,U) = \beta_0 + \beta_a A + \beta_u U + \beta_x^{\mathrm{T}} X,$$
$$\mathbb{E}(W|A,X,V,U) = \gamma_0 + \gamma_u U + \gamma_x^{\mathrm{T}} X,$$

noting the DAG in Figure 6.3 implies $Y \perp\!\!\!\perp V|(A,X,U)$ and $W \perp\!\!\!\perp (A,V)|(X,U)$.

Because $Y \perp\!\!\!\perp A|(X,U)$, with the help of SCMs, we can show that the following condition, which is similar to (6.7), is satisfied,

$$(Y^{a=0}, Y^{a=1}) \perp\!\!\!\perp A|(X,U).$$

Thus we can express θ^*_{ATE} in terms of coefficient β_a,

$$\begin{aligned}
\theta^*_{\mathrm{ATE}} &= \mathbb{E}[\mathbb{E}(Y|A=1,X,U)] - \mathbb{E}[\mathbb{E}(Y|A=0,X,U)] \\
&= \mathbb{E}[\beta_0 + \beta_a(1) + \beta_u U + \beta_x^{\mathrm{T}} X] - \mathbb{E}[\beta_0 + \beta_a(0) + \beta_u U + \beta_x^{\mathrm{T}} X] \\
&= \beta_a.
\end{aligned}$$

To see that β_a is identifiable, we apply the law of iterated expectations to the above structural linear models and obtain that

$$\mathbb{E}(Y|A,X,V) = \beta_0 + \beta_a A + \beta_u \mathbb{E}(U|A,X,V) + \beta_x^{\mathrm{T}} X,$$
$$\mathbb{E}(W|A,X,V) = \gamma_0 + \gamma_u \mathbb{E}(U|A,X,V) + \gamma_x^{\mathrm{T}} X.$$

Thus, if $\gamma_u \neq 0$, we have

$$\mathbb{E}(Y|A,X,V) = \widetilde{\beta}_0 + \beta_a A + \widetilde{\beta}_u \mathbb{E}(W|A,X,V) + \widetilde{\beta}_x^{\mathrm{T}} X,$$

where

$$\widetilde{\beta}_0 = \beta_0 - \beta_u \gamma_0 / \gamma_u,$$
$$\widetilde{\beta}_u = \beta_u / \gamma_u,$$
$$\widetilde{\beta}_x = \beta_x - \beta_u \gamma_x / \gamma_u.$$

Since $\mathbb{E}(Y|A, X, V)$ and $\mathbb{E}(W|A, X, V)$ are identifiable, β_a is identifiable. Therefore, we demonstrate that θ^*_{ATE} is identifiable with the help of structural linear models. Miao, Geng, and Tchetgen Tchetgen (2018) demonstrated that θ^*_{ATE} is non-parametrically identifiable under a certain rank assumption.

Tchetgen Tchetgen et al. (2020) referred to the causal inference in Step 1 as classical causal inference and the causal inference in Step 3 as proximal causal inference. Proximal causal inference is an emerging causal inference framework. In this book, we focus on the classical causal inference framework.

6.3 Longitudinal Cohort Studies

6.3.1 Causal estimand

As in Chapter 5, let $\overline{A}(t) = (A(0), \ldots, A(t))$ be the observed treatment sequence up to t, and let $\overline{X}(t) = (X(0), \ldots, X(t))$ be the vector consisting of all the observed history up to time t including baseline covariates, time-dependent covariates, and intermediate outcomes, $t = 0, \ldots, T-1$. In particular, define $\overline{A} = (A(0), \ldots, A(T-1))$ and $\overline{X} = (X(0), \ldots, X(T-1))$. Let Y be the outcome variable measured at time T.

Let $Y^{\overline{a}}$ be the potential outcome had the patient been treated by treatment sequence $\overline{a} = (a(0), a(1), \ldots, a(T-1))$. Let $\overline{a}(t) = (a(0), a(1), \ldots, a(t))$. Two particular treatment sequences are $\overline{a} = \overline{1} = (1, \ldots, 1)$ and $\overline{a} = \overline{0} = (0, \ldots, 0)$. As in Chapter 5, define the set of all potential outcomes associated with $\overline{a}$,

$$W^{\overline{a}} = \left\{ X^{a(0)}(1), X^{\overline{a}(t-1)}(t), t = 2, \ldots, T-1, Y^{\overline{a}} \right\}. \tag{6.8}$$

After defining the above potential outcomes, we define the causal estimand of interest. For example, we are interested in the following causal estimand,

$$\theta^*_{\text{ATE}} = \mathbb{E}\left(Y^{\overline{a}=\overline{1}}\right) - \mathbb{E}\left(Y^{\overline{a}=\overline{0}}\right), \tag{6.9}$$

where the expectation is over the population. It measures the difference between the mean of the potential outcomes if all the patients in the population had been treated by treatment sequence $\overline{a} = \overline{1}$ and that if all the patients in the population had been treated by treatment sequence $\overline{a} = \overline{0}$.

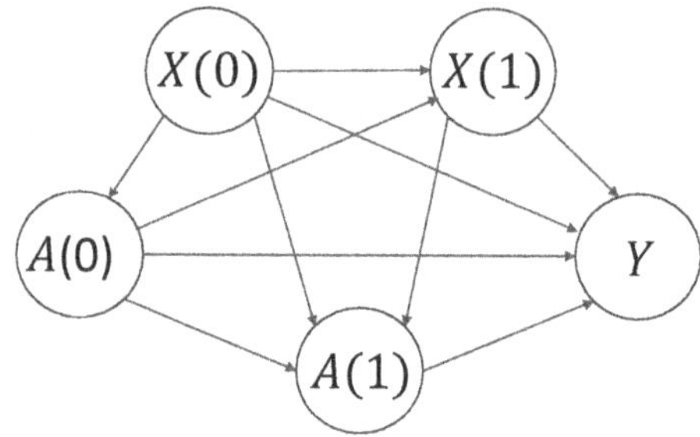

FIGURE 6.4
A longitudinal cohort study with $T = 2$

6.3.2 Identifiability assumptions

For the goal of identification, we make three identifiability assumptions:

1. The consistency assumption;
2. The assumptions embedded in the DAG—like the one in Figure 6.4;
3. The positivity assumption.

Let's understand the consistency assumption first. Besides the potential (primary) outcome $Y^{\overline{a}}$, there are potential (intermediate) outcomes $X^{a(0)}(1)$ and $X^{\overline{a}(t)}(t)$, $t = 2, \ldots, T-1$. Thus, we assume the following,

1. The consistency assumption:

$$\begin{aligned} X^{a(0)}(1) &= X(1), \text{ if } A(0) = a(0); \\ X^{\overline{a}(t-1)}(t) &= X(t), \text{ if } \overline{A}(t-1) = \overline{a}(t-1), t = 2, \ldots, T-1; \\ Y^{\overline{a}} &= Y, \text{ if } \overline{A} = \overline{a}. \end{aligned}$$

Next, let's see what assumptions are embedded in the DAG. With the help of SCMs (see Appendix 6.5 for more detail), we can show that at time t, given the history up to t, that is given $(\overline{A}(t-1), \overline{X}(t))$, the treatment $A(t)$ is independent of the potential outcomes. Thus, the DAG implies the following,

2. The sequential exchangeability assumption:

$$\begin{aligned} &W^{\overline{a}} \perp\!\!\!\perp A(0) | X(0), \\ &W^{\overline{a}} \perp\!\!\!\perp A(t) | (\overline{A}(t-1), \overline{X}(t)). \end{aligned}$$

Last, let's state the positivity assumption. The positivity assumption depends on which treatment sequences are considered. If the causal estimand is the treatment effect of $\overline{a} = \overline{1}$ vs. $\overline{a} = \overline{0}$, we assume the following,

3. The positivity assumption:

$$\mathbb{P}(\overline{A} = \overline{j}|X(0) = x) > 0, \text{ for } j = 1, 0, x \in \text{supp}(X(0)).$$

6.3.3 Identification

Under the above three identifiability assumptions, we are able to accomplish the task of identification. And there are two strategies: the standardization strategy and the weighting strategy (Fang 2020; Tsiatis et al. 2020).

The standardization strategy

For the purpose of demonstration, we use probability mass functions for all variables in $X(t)$ and Y. For $T = 2$, we have

$$\begin{aligned}
&\mathbb{P}\{Y = y|X(0) = x(0), A(0) = a(0), X(1) = x(1), A(1) = a(1)\}\\
\overset{(a)}{=}&\mathbb{P}\{Y^{\overline{a}} = y|X(0) = x(0), A(0) = a(0), X(1) = x(1), A(1) = a(1)\}\\
\overset{(b)}{=}&\mathbb{P}\{Y^{\overline{a}} = y|X(0) = x(0), A(0) = a(0), X(1) = x(1)\}\\
\overset{(c)}{=}&\mathbb{P}\{Y^{\overline{a}} = y|X(0) = x(0), A(0) = a(0), X^{a(0)}(1) = x(1)\}\\
\overset{(d)}{=}&\frac{\mathbb{P}\{Y^{\overline{a}} = y, X^{a(0)}(1) = x(1)|X(0) = x(0), A(0) = a(0)\}}{\mathbb{P}\{X^{a(0)}(1) = x(1)|X(0) = x(0), A(0) = a(0)\}}\\
\overset{(e)}{=}&\frac{\mathbb{P}\{Y^{\overline{a}} = y, X^{a(0)}(1) = x(1)|X(0) = x(0)\}}{\mathbb{P}\{X^{a(0)}(1) = x(1)|X(0) = x(0)\}}\\
\overset{(f)}{=}&\mathbb{P}\{Y^{\overline{a}} = y|X(0) = x(0), X^{a(0)}(1) = x(1)\},
\end{aligned}$$

where (a) holds under the consistency assumption, (b) holds under the sequential exchangeability assumption, (c) holds under the consistency assumption again, (d) holds using the relationship among joint, marginal, and conditional distributions, (e) holds applying the sequential exchangeability assumption in both the numerator and denominator, and (f) holds using the relationship among joint, marginal, and conditional distributions.

We also have

$$\begin{aligned}
&\mathbb{P}\{X(1) = x(1)|X(0) = x(0), A(0) = a(0)\}\\
\overset{(a)}{=}&\mathbb{P}\{X^{a(0)}(1) = x(1)|X(0) = x(0), A(0) = a(0)\}\\
\overset{(b)}{=}&\mathbb{P}\{X^{a(0)}(1) = x(1)|X(0) = x(0)\},
\end{aligned}$$

where (a) holds under the consistency assumption and (b) holds using the relationship among joint, marginal, and conditional distributions again.

Combining the above two results, we have

$$
\begin{aligned}
&\mathbb{P}\{X(0) = x(0), X^{a(0)}(1) = x(1), Y^{\overline{a}} = y\} \\
=&\mathbb{P}\{X(0) = x(0)\} \times \mathbb{P}\{X^{a(0)}(1) = x(1)|X(0) = x(0)\} \\
&\times \mathbb{P}\{Y^{\overline{a}} = y|X(0) = x(0), X^{a(0)}(1) = x(1)\} \\
=&\mathbb{P}\{X(0) = x(0)\} \times \mathbb{P}\{X(1) = x(1)|X(0) = x(0), A(0) = a(0)\} \\
&\times \mathbb{P}\{Y = y|X(0) = x(0), A(0) = a(0), X(1) = x(1), A(1) = a(1)\}.
\end{aligned}
$$

Therefore, for $T = 2$, we show that

$$
\mathbb{E}\left(Y^{\overline{a}}\right) = \mathbb{E}_{\overline{X} \sim \mathcal{F}_{\overline{a}}}\left[\mathbb{E}(Y|\overline{X}, \overline{A} = \overline{a})\right], \tag{6.10}
$$

where $\mathcal{F}_{\overline{a}}$ is the following distribution function of $\overline{X}$,

$$
\begin{aligned}
&\mathbb{P}_{\mathcal{F}_{\overline{a}}}(X(0) = x(0), \ldots, X(T-1) = x(T-1)) \\
=&\mathbb{P}\{X(0) = x(0)\} \times \mathbb{P}\{X(1) = x(1)|X(0) = x(0), A(0) = a(0)\}. \ \square
\end{aligned}
$$

Thus, under the identifiability assumptions, we translate the causal estimand into the following statistical estimand:

$$
\theta_{\text{ATE}} = \mathbb{E}_{\overline{X} \sim \mathcal{F}_{\overline{1}}}\left[\mathbb{E}(Y|\overline{X}, \overline{A} = \overline{1})\right] - \mathbb{E}_{\overline{X} \sim \mathcal{F}_{\overline{0}}}\left[\mathbb{E}(Y|\overline{X}, \overline{A} = \overline{0})\right]. \tag{6.11}
$$

The weighting strategy

Also start with $T = 2$. We have

$$
\begin{aligned}
&\mathbb{E}\left[\frac{I(\overline{A} = \overline{a})Y}{\mathbb{P}\{A(0) = a(0)|X(0)\}\mathbb{P}\{A(1) = a(1)|X(0), A(0), X(1)\}}\right] \\
\overset{(a)}{=}&\mathbb{E}\left[\frac{I(\overline{A} = \overline{a})Y^{\overline{a}}}{\mathbb{P}\{A(0) = a(0)|X(0)\}\mathbb{P}\{A(1) = a(1)|X(0), A(0), X^{a(0)}(1)\}}\right] \\
\overset{(b)}{=}&\mathbb{E}\left[\frac{\mathbb{E}\{I(\overline{A} = \overline{a})Y^{\overline{a}}|X(0), W^{\overline{a}}\}}{\mathbb{P}\{A(0) = a(0)|X(0)\}\mathbb{P}\{A(1) = a(1)|X(0), A(0) = a(0), X^{a(0)}(1)\}}\right] \\
\overset{(c)}{=}&\mathbb{E}\left[\frac{\mathbb{P}\{\overline{A} = \overline{a}|X(0), W^{\overline{a}}\}Y^{\overline{a}}}{\mathbb{P}\{A(0) = a(0)|X(0)\}\mathbb{P}\{A(1) = a(1)|X(0), A(0) = a(0), X^{a(0)}(1)\}}\right] \\
\overset{(d)}{=}&\mathbb{E}\left[\frac{\mathbb{P}\{A(0) = a(0)|X(0), W^{\overline{a}}\}\mathbb{P}\{A(1) = a(1)|X(0), A(0) = a(0), W^{\overline{a}}\}Y^{\overline{a}}}{\mathbb{P}\{A(0) = a(0)|X(0)\}\mathbb{P}\{A(1) = a(1)|X(0), A(0) = a(0), X^{a(0)}(1)\}}\right] \\
\overset{(e)}{=}&\mathbb{E}(Y^{\overline{a}}),
\end{aligned}
$$

where the left-hand-side of (a) is well-defined under the positivity assumption, the right-hand-side of (a) holds under the consistency assumption, (b) holds using the law of iterated expectations, (c) holds because the expectation of the indicator of an event is the probability of the event, (d) holds using the relationship among joint, marginal, and conditional distributions, and (e) holds under the sequential exchangeability assumption.

Thus, under the three identifiability assumptions, using the weighting strategy, we define the following statistical estimand:

$$\theta_{\text{ATE}} = \mathbb{E}\left[\frac{I(\overline{A}=\overline{1})Y}{\mathbb{P}\{A(0)=1|X(0)\}\prod_{t=1}^{T-1}\mathbb{P}\{A(t)=1|\overline{A}(t-1)=\overline{1},\overline{X}(t)\}}\right]$$
$$-\mathbb{E}\left[\frac{I(\overline{A}=\overline{0})Y}{\mathbb{P}\{A(0)=0|X(0)\}\prod_{t=1}^{T-1}\mathbb{P}\{A(t)=0|\overline{A}(t-1)=\overline{0},\overline{X}(t)\}}\right]. \quad (6.12)$$

Note the above two statistical estimands, (6.11) and (6.12), are equal to each other under the three identifiability assumptions.

6.4 Externally Controlled Trials

Consider an ECT (i.e., a single-arm clinical trial with an external control group). Let $A = 1$ indicate that the subject is in the single-arm trial (the treated arm) and $A = 0$ indicate that the subject is in the external control group (the control arm). If the external controls are obtained from the placebo arms of some historical RCTs, then the treatment in the control arm is placebo; if the external controls are obtained from RWD, then the treatment in the control arm may be the standard of care. Therefore, we can redefine A as the treatment variable, with 1 standing for the investigative treatment and 0 the control treatment.

Let n_1 and n_0 be the sample sizes of the treated arm and the control arm, respectively. Let $n = n_1 + n_0$ be the total sample size. Let X be the vector of pre-treatment variables and Y be the outcome variable of interest.

6.4.1 Causal estimand

Let $Y^{a=j}$ be the potential outcome that one patient would have had if the patient had been treated by treatment $j = 1$ or 0.

In Section 6.2, we discuss the relationship between sample and population in a cohort study. Let's practice this mental exercise again for ECT. In an ECT, the target population is the one from which n_1 subjects are drawn to participate in the single-arm clinical trial. We refer to this conceptually constructed population as the treated population. Thus, in an ECT, we are interested in the following causal estimand:

$$\theta^*_{\text{ATT}} = \mathbb{E}(Y^{a=1}|A=1) - \mathbb{E}(Y^{a=0}|A=1), \quad (6.13)$$

where the expectation is over the treated population. As suggested by the subscript, this causal estimand is referred to as the average treatment effect among the treated (ATT), which measures the difference between the mean of the outcomes in the treated population and that if all the patients in the treated population—who are treated by treatment 1—had been treated by treatment 0 counterfactually.

6.4.2 Identification

We assume the same DAG in Figure 6.1. For the purpose of identification, we make the same identifiability assumptions as those made in Section 6.2 (consistency, exchangeability, and positivity). Similarly, there are two strategies, the standardization strategy and the weighting strategy, to accomplish the task the identification.

The standardization strategy

Under those three identifiability assumptions, we can show that

$$\begin{aligned}
\theta^*_{\text{ATT}} &= \mathbb{E}(Y^{a=1}|A=1) - \mathbb{E}(Y^{a=0}|A=1) \\
&\overset{(a)}{=} \mathbb{E}(Y|A=1) - \mathbb{E}(Y^{a=0}|A=1) \\
&\overset{(b)}{=} \mathbb{E}(Y|A=1) - \mathbb{E}_{X|A=1}\left[\mathbb{E}(Y^{a=0}|X, A=1)\right] \\
&\overset{(c)}{=} \mathbb{E}(Y|A=1) - \mathbb{E}_{X|A=1}\left[\mathbb{E}(Y^{a=0}|X, A=0)\right] \\
&\overset{(d)}{=} \mathbb{E}(Y|A=1) - \mathbb{E}_{X|A=1}\left[\mathbb{E}(Y|X, A=0)\right],
\end{aligned}$$

where (a) holds under the consistency assumption, (b) holds using the law of iterated expectations and "$\mathbb{E}_{X|A=1}$" means the expectation is over the conditional distribution of X given $A = 1$, (c) holds under the exchangeability and positivity assumptions, and (d) holds under the consistency assumption.

Thus, under those three assumptions, using the standardization strategy, we translate the causal estimand into the following statistical estimand,

$$\theta_{\text{ATT}} = \mathbb{E}(Y|A=1) - \mathbb{E}_{X|A=1}\left[\mathbb{E}(Y|X, A=0)\right]. \tag{6.14}$$

The weighting strategy

Under those three identifiability assumptions, we can show that

$$\begin{aligned}
&\mathbb{E}\left[\frac{I(A=0)\mathbb{P}(A=1|X)Y}{\mathbb{P}(A=1)\mathbb{P}(A=0|X)}\right] \\
&\overset{(a)}{=}\mathbb{E}\left[\frac{I(A=0)\mathbb{P}(A=1|X)Y^{a=0}}{\mathbb{P}(A=1)\mathbb{P}(A=0|X)}\right] \\
&\overset{(b)}{=}\mathbb{E}\left\{\mathbb{E}\left[\frac{I(A=0)\mathbb{P}(A=1|X)Y^{a=0}}{\mathbb{P}(A=1)\mathbb{P}(A=0|X)}\middle| X\right]\right\} \\
&\overset{(c)}{=}\mathbb{E}\left\{\frac{\mathbb{P}(A=1|X)\mathbb{E}[Y^{a=0}|X]}{\mathbb{P}(A=1)}\right\} \\
&\overset{(d)}{=}\mathbb{E}_{X|A=1}\left\{\mathbb{E}[Y^{a=0}|X, A=1]\right\} \\
&\overset{(e)}{=}\mathbb{E}[Y^{a=0}|A=1],
\end{aligned}$$

where the left-hand-side of (a) is well-defined under the positivity assumption, the right-hand-side of (a) holds under the consistency assumption, (b) holds using the law of iterated expectations, (c) holds under the exchangeability assumption, (d) holds using Bayes' theorem, and (e) holds using the law of iterated expectations.

Thus, under those three assumptions, using the weighting strategy, we translate the causal estimand into the following statistical estimand,

$$\theta_{\text{ATT}} = \mathbb{E}\left[\frac{I(A=1)Y}{\mathbb{P}(A=1)}\right] - \mathbb{E}\left[\frac{I(A=0)\mathbb{P}(A=1|X)Y}{\mathbb{P}(A=1)\mathbb{P}(A=0|X)}\right]. \tag{6.15}$$

Note the above two statistical estimands, (6.14) and (6.15), are equal to each other under the identifiability assumptions.

We conclude this section with some discussion on the average treatment effect on the controls (ATC),

$$\theta^*_{\text{ATC}} = \mathbb{E}(Y^{a=1}|A=0) - \mathbb{E}(Y^{a=0}|A=0), \tag{6.16}$$

where the population is the one from which the controls are drawn. Thus, all the above discussion still applies except that we exchange $A = 1$ and $A = 0$.

6.5 Appendix

SCMs for DAG in Figure 6.1

We can write down the relationships among X, A, and Y by a series of SCMs,

$$\begin{aligned} X &= f_X(U_X), \\ A &= f_A(X, U_A), \\ Y &= f_Y(X, A, U_Y), \end{aligned}$$

where f_X, f_A, and f_Y are unknown deterministic functions and U_X, U_A, and U_Y are independent.

Based on the SCMs, we can express the potential outcomes explicitly as

$$\begin{aligned} Y^{a=1} &= f_Y(X, 1, U_Y), \\ Y^{a=0} &= f_Y(X, 0, U_Y). \end{aligned}$$

Because $U_Y \perp\!\!\!\perp U_A | X$, we see that

$$(f_Y(X, 1, U_Y), Y_Y(X, 0, U_Y)) \perp\!\!\!\perp f_A(X, U_A) | X,$$

which is equivalent to

$$(Y^{a=1}, Y^{a=0}) \perp\!\!\!\perp A | X.$$

SCMs for DAG in Figure 6.2

We can write down the relationships among X, A, and Y by a series of SCMs,

$$\begin{aligned} X &= f_X(U_X), \\ U &= f_U(U_U), \\ A &= f_A(X, U, U_A), \\ Y &= f_Y(X, U, A, U_Y), \end{aligned}$$

where f_X, f_U, f_A, and f_Y are unknown deterministic functions and U_X, U_U, U_A, and U_Y are independent.

Based on the SCMs, we can express the potential outcomes explicitly as

$$\begin{aligned} Y^{a=1} &= f_Y(X, U, 1, U_Y), \\ Y^{a=0} &= f_Y(X, U, 0, U_Y). \end{aligned}$$

Because both sides involve U, we see that

$$(f_Y(X, U, 1, U_Y), Y_Y(X, U, 0, U_Y)) \not\perp\!\!\!\perp f_A(X, U, U_A)|X,$$

which is equivalent to

$$(Y^{a=1}, Y^{a=0}) \not\perp\!\!\!\perp A|X.$$

But, because $U_Y \perp\!\!\!\perp U_A|(X, U)$, we see that

$$(f_Y(X, U, 1, U_Y), Y_Y(X, U, 0, U_Y)) \not\perp\!\!\!\perp f_A(X, U, U_A)|(X, U),$$

which is equivalent to

$$(Y^{a=1}, Y^{a=0}) \perp\!\!\!\perp A|(X, U).$$

SCMs for a longitudinal study

The data generaling process for a longitudinal study can be expressed by a series of SCMs:

$$\begin{aligned} X(0) &= f_{X(0)}(U_{X(0)}), \\ A(0) &= f_{A(0)}(X(0), U_{A(0)}), \\ X(1) &= f_{X(1)}(X(0), A(0), U_{X(1)}), \\ A(1) &= f_{A(1)}(X(0), A(0), X(1), U_{A(1)}), \\ X(t) &= f_{X(t)}(X(0), A(0), \ldots, X(t-1), A(t-1), U_{X(t)}), \\ A(t) &= f_{A(t)}(X(0), A(0), \ldots, A(t-1), X(t), U_{A(t)}), t = 2, \ldots, T-1, \\ Y &= f_Y(X(0), A(0), \ldots, X(T-1), A(T-1), U_Y), \end{aligned}$$

where deterministic functions $f.$'s are unknown and exogenous variables $U.$'s are independent.

We can express potential outcomes explicitly:

$$\begin{aligned}X^{a(0)}(1) &= f_{X(1)}(X(0), a(0), U_{X(1)}),\\ X^{\overline{a}(t)}(t) &= f_{X(t)}(X(0), a(0), \ldots, X^{\overline{a}(t-2)}(t-1), a(t-1), U_{X(t)}),\\ &\qquad t = 2, \ldots, T-1,\\ Y^{\overline{a}} &= f_Y(X(0), a(0), \ldots, X^{\overline{a}(T-2)}(T-1), a(T-1), U_Y).\end{aligned}$$

And recall that

$$W^{\overline{a}} = \{X^{a(0)}(1), X^{\overline{a}(t-1)}(t), t = 2, \ldots, T-1, Y^{\overline{a}}\}.$$

Because $U_{A(0)} \perp\!\!\!\perp \{U_{X(t)}, t = 1, \ldots, T-1; U_Y\} | X(0)$, we see that

$$f_{A(0)}(X(0), U_{A(0)} \perp\!\!\!\perp W^{\overline{a}} | X(0),$$

which is equivalent to

$$A(0) \perp\!\!\!\perp W^{\overline{a}} | X(0).$$

Similarly, because $U_{A(t)} \perp\!\!\!\perp \{U_{X(t)}, t = 1, \ldots, T-1; U_Y\} | (\overline{A}(t-1), \overline{X}(t))$, we see that

$$f_{A(t)}(\overline{A}(t-1), \overline{X}(t), U_{A(t)}) \perp\!\!\!\perp W^{\overline{a}} | (\overline{A}(t-1), \overline{X}(t)),$$

which is equivalent to

$$A(t) \perp\!\!\!\perp W^{\overline{a}} | (\overline{A}(t-1), \overline{X}(t)).$$

6.6 Exercises

Ex 6.1

In a cohort study whose DAG is displayed in Figure 6.1, there are analysis dropouts indicated by Δ. Let $\Delta = 1$ and 0 indicate that Y is missing and observed, respectively. We are interested in the ATE causal estimand, θ^*_{ATE} defined in (6.1). In addition to the three identifiability assumptions made in Section 6.2, we make the following two assumptions:

4. The MAR assumption:
$$(Y^{a=0}, Y^{a=1}) \perp\!\!\!\perp \Delta | X.$$

5. An additional positivity assumption:
$$\mathbb{P}(\Delta = 0 | X = x) > 0, \text{ for } x \in \text{supp}(X).$$

Show that θ^*_{ATE} is identifiable using the standardization strategy.

Ex 6.2

Continue Ex 6.1. Show that θ^*_{ATE} is identifiable using the weighting strategy.

Ex 6.3

Continue Ex 6.1. Assume that Y is a 0/1–binary variable, with 1 standing for failure and 0 for success. Consider analysis dropout Δ as ICE and consider the composite variable strategy—one of the five ICE handling strategies in ICH E9(R1)—to deal with analysis dropouts. That is, we define the following new outcome variable:

$$\widetilde{Y} = 1 \text{ if } \; Y = 1 \text{ or } \; \Delta = 1,$$
$$\widetilde{Y} = 0 \text{ if } \; Y = 0 \text{ and } \; \Delta = 0.$$

By the composite-variable strategy, we define the following causal estimand,

$$\widetilde{\theta}^*_{\text{ATE}} = \mathbb{E}(\widetilde{Y}^{a=1}) - \mathbb{E}(\widetilde{Y}^{a=0}),$$

where $\widetilde{Y}^{a=j}$ is the potential outcome of $\widetilde{Y}$ if the patient had been treated by treatment j, $j = 1, 0$. Write down the three identifiability assumptions under which $\widetilde{\theta}^*_{\text{ATE}}$ is identifiable.

7

The Art of Estimation (I): M-estimation

7.1 Introduction

Continue the discussion in Chapter 6. Consider a cohort study, in which we are interested in estimating the average treatment effect (ATE),

$$\theta^* = \mathbb{E}(Y^{a=1}) - \mathbb{E}(Y^{a=0}), \tag{7.1}$$

based on the observed dataset $\{O_i = (X_i, A_i, Y_i), i = 1, \ldots, n\}$.

In Chapter 6, we showed that under the identifiability assumptions—consistency, exchangeability, and positivity—we can translate the causal estimand θ^* to a statistical estimand θ and there are two strategies to do so: the standardization strategy and the weighting strategy.

By the standardization strategy, we show that θ^* is equal to

$$\theta = \mathbb{E}[\mathbb{E}(Y|A = 1, X)] - \mathbb{E}[\mathbb{E}(Y|A = 0, X)]. \tag{7.2}$$

The inner expectations, $\mathbb{E}(Y|A = a, X)$, $a = 1, 0$, can be seen as the regression function in modeling of outcome Y against treatment variable A and covariates X; that is, $Y \sim A + X$. Thus, we define the following function,

$$Q(x, a) = \mathbb{E}(Y|A = a, X = x) \tag{7.3}$$

and refer to it as the *outcome regression function.*

Alternatively, by the weighting strategy, we show that θ^* is equal to

$$\theta = \mathbb{E}\left[\frac{I(A = 1)}{\mathbb{P}(A = 1|X)}Y\right] - \mathbb{E}\left[\frac{I(A = 0)}{\mathbb{P}(A = 0|X)}Y\right]. \tag{7.4}$$

The denominators inside the expectations, $\mathbb{P}(A = a|X)$, $a = 1, 0$, are called the propensity score in modeling treatment variable A against covariates X; that is, $A \sim X$. Thus, we define the following function,

$$g(a|x) = \mathbb{P}(A = a|X = x) \tag{7.5}$$

and refer to it as the *propensity score function.*

DOI: 10.1201/9781003433378-7

TABLE 7.1
A preview of four estimation methods

	Weighting	Standardization
Initial estimator	IPW	MLE
Doubly robust estimator	AIPW	TMLE

The above two versions of statistical estimand, (7.2) via the standardization strategy and (7.4) via the weighting strategy, are equivalent because both of them are equal to the causal estimand θ^* under the same identifiability assumptions. Moving forward, these two versions of the statistical estimand will lead to two major classes of estimators: (i) a class of estimators aiming for estimating (7.2) and (ii) a class of estimators aiming for estimating (7.4).

In this chapter, we will describe the M-estimation framework (Hernán and Robins 2020), and in the next two chapters, we will describe the targeted learning framework (van der Laan and Rose 2011). Together, we will describe two representatives from Class (i), the inverse probability weighted estimator (IPW) and the augmented inverse probability weighted estimator (AIPW), and two representatives from Class (ii): the maximum likelihood estimator or minimum loss estimator (MLE) and the targeted maximum likelihood estimator or the targeted minimum loss estimator (TMLE). These four estimators are previewed in Table 7.1.

7.2 M-estimation

7.2.1 M-estimator

The M-estimation method was first proposed by Huber (1964) and has been widely studied; e.g., van der Vaart (2000).

Consider a parametric model that generates i.i.d. observations $O_1, \ldots, O_n$, with $O \sim f_O(o; \epsilon)$, where ϵ is a parameter specifying the distribution. Assume that we are interested in estimating the q-dimensional parameter $\mu = \mu(\epsilon)$. Assume ϵ_0 is the true value of the parameter and $\mu_0 = \theta(\epsilon_0)$. We want to construct an estimator for μ and estimate its asymptotic variance. An M-estimator for μ, denoted as $\widehat{\mu}$, is the solution of the estimating equation,

$$\sum_{i=1}^{n} M(O_i; \widehat{\mu}) = 0, \tag{7.6}$$

where $M(o; \mu) = \{M_1(o; \mu), \ldots, M_q(o; \mu)\}^{\mathrm{T}}$ is a $q \times 1$ estimating function satisfying

$$\mathbb{E}\{M(O; \mu_0)\} = 0, \tag{7.7}$$

where the expectation is taken over the true distribution $f_O(o;\epsilon_0)$.

Consider the first-order Taylor expansion of (7.6) in $\widehat{\mu}$ about μ_0:

$$0=\sum_{i=1}^{n} M(O_i;\mu_0)+\left\{\sum_{i=1}^{n}\frac{\partial M(O_i;\mu_0)}{\partial\mu^{\mathrm{T}}}\right\}(\widehat{\mu}-\mu_0)+o_p(\sqrt{n}).$$

We need some regularity conditions to ensure the remainder term divided by $\sqrt{n}$ is $o_p(1)$, which is the notation for a term converging to zero in probability. Here we omit theoretical arguments and focus on heuristic arguments. By the law of large numbers, we have

$$\frac{1}{n}\sum_{i=1}^{n}\frac{\partial M(O_i;\mu_0)}{\partial\mu^{\mathrm{T}}}\xrightarrow{p}\mathbb{E}\left[\frac{\partial M(O,\mu_0)}{\partial\mu^{\mathrm{T}}}\right].$$

Thus, by Slutsky's lemma, we have

$$\sqrt{n}(\widehat{\mu}-\mu_0)=-\left\{\mathbb{E}\left[\frac{\partial M(O;\mu_0)}{\partial\mu^{\mathrm{T}}}\right]\right\}^{-1}\frac{1}{\sqrt{n}}\sum_{i=1}^{n} M(O_i;\mu_0)+o_p(1). \quad (7.8)$$

By the central limit theorem,

$$\frac{1}{\sqrt{n}}\sum_{i=1}^{n} M(O_i;\mu_0)\xrightarrow{d}\mathcal{N}\left(0,\mathbb{E}\left\{M^{\otimes 2}(O;\mu_0)\right\}\right),$$

where $M^{\otimes 2}(O;\mu_0)=M(O;\mu_0)M^{\mathrm{T}}(0;\mu_0)$. Thus, we heuristically show that

$$\widehat{\mu}\xrightarrow{p}\mu_0, \quad (7.9)$$

$$\sqrt{n}(\widehat{\mu}-\mu_0)\xrightarrow{d}\mathcal{N}(0,\Sigma), \quad (7.10)$$

where

$$\Sigma=\left\{\mathbb{E}\left[\frac{\partial M(O;\mu_0)}{\partial\mu^{\mathrm{T}}}\right]\right\}^{-1}\mathbb{E}\left\{M^{\otimes 2}(O;\mu_0)\right\}\left\{\mathbb{E}\left[\frac{\partial M(O;\mu_0)}{\partial\mu^{\mathrm{T}}}\right]\right\}^{-\mathrm{T}}, \quad (7.11)$$

which is often referred to as *the sandwich formula.*

In asymptotic statistics, the property expressed in (7.9) is called the consistency (not to be confused with the consistency assumption), and the property expressed in (7.10) is called the asymptotic normality.

In the sandwich formula, the bread term can be estimated by

$$\widehat{\mathbb{E}}\left[\frac{\partial M(O;\mu_0)}{\partial\mu^{\mathrm{T}}}\right]=\frac{1}{n}\sum_{i=1}^{n}\frac{\partial M(O_i;\widehat{\mu})}{\partial\mu^{\mathrm{T}}},$$

and the meat term can be estimated by

$$\widehat{\mathbb{E}}\left\{M^{\otimes 2}(O;\mu_0)\right\}=\frac{1}{n}\sum_{i=1}^{n} M^{\otimes 2}(O_i;\widehat{\mu}).$$

Combining the bread term and the meat term, Σ can be estimated by

$$\widehat{\Sigma} = n\left\{\sum_{i=1}^{n} \frac{\partial M(O_i;\widehat{\mu})}{\partial \mu^{\mathrm{T}}}\right\}^{-1}\left[\sum_{i=1}^{n} M^{\otimes 2}(O_i;\widehat{\mu})\right]\left\{\sum_{i=1}^{n} \frac{\partial M(O_i;\widehat{\mu})}{\partial \mu^{\mathrm{T}}}\right\}^{-\mathrm{T}}.$$

7.2.2 Asymptotic linearity

In asymptotic statistics, we call an estimator of form (7.8) as an asymptotically linear estimator, because it can be rewritten as

$$\sqrt{n}(\widehat{\mu} - \mu_0) = \frac{1}{\sqrt{n}}\sum_{i=1}^{n} \phi(O_i) + o_p(1), \tag{7.12}$$

where

$$\phi(O) = -\left\{\mathbb{E}\left[\frac{\partial M(O;\mu_0)}{\partial \mu^{\mathrm{T}}}\right]\right\}^{-1} M(O;\mu_0) \tag{7.13}$$

is called the influence function (IF). Note that

$$\mathbb{E}\{\phi(O)\} = 0,$$
$$\mathbb{E}\{\phi(O)\phi^{\mathrm{T}}(O)\} = \Sigma.$$

"The name 'influence function' originated in developing robust statistics. The function measures the change in the value $\phi(P)$ if an infinitesimally small part of P is replaced by a pointmass at x."—van der Vaart (2000)

7.2.3 Regularity

The theory of likelihood

To provide a brief review of the concepts of likelihood, score, Fisher information, and Cramer-Rao bound, we consider a parametric setting where $O_1, \ldots, O_n$ i.i.d. with $f(o;\epsilon)$, where ϵ is a scalar parameter.

Likelihood, log-likelihood, and the score function are

$$L(\epsilon) = \prod_{i=1}^{n} f(O_i;\epsilon),$$

$$l(\epsilon) = \log L(\epsilon) = \sum_{i=1}^{n} \log f(O_i;\epsilon),$$

$$S_n(\epsilon) = \partial l(\epsilon)/\partial \epsilon = \sum_{i=1}^{n} \frac{\partial \log f(O_i;\epsilon)}{\partial \epsilon},$$

respectively. MLE is equivalent to the solution of the score function; i.e.,

$$\widehat{\epsilon}_{\text{MLE}} = \arg\max_{\epsilon} L(\epsilon) \Longleftrightarrow S_n(\widehat{\epsilon}_{\text{MLE}}) = 0. \tag{7.14}$$

Let ϵ_0 be the true value of ϵ. Define

$$S_\epsilon = \partial \log f(O; \epsilon_0)/\partial\epsilon,$$

and refer it as the *score*. We can show that

$$\mathbb{E}(S_\epsilon) = 0. \tag{7.15}$$

In fact, taking the derivative over ϵ on both sides of $\int f(o;\epsilon)do = 1$, we have $\int \partial f(o;\epsilon)/\partial\epsilon do = 0$. Hence, we have

$$\int \frac{\partial f(o;\epsilon_0)/\partial\epsilon}{f(o;\epsilon_0)} f(o;\epsilon_0)do = \mathbb{E}\left[\frac{\partial \log f(o;\epsilon_0)}{\partial\epsilon}\right] = \mathbb{E}(S_\epsilon) = 0.$$

Thus, based on (7.14), by the M-estimation theory, we can show that $\widehat{\epsilon}_{\text{MLE}}$ is consistent and asymptotically linear,

$$\sqrt{n}(\widehat{\epsilon}_{\text{MLE}} - \epsilon_0) = -\frac{1}{\sqrt{n}}\sum_{i=1}^{n}\left\{\mathbb{E}\left[\frac{\partial^2 \log f(O;\epsilon_0)}{\partial\epsilon^2}\right]\right\}^{-1}\frac{\partial \log f(O_i;\epsilon_0)}{\partial\epsilon} + o_p(1).$$

Moreover, we can show that

$$-\mathbb{E}\left[\frac{\partial^2 \log f(O;\epsilon_0)}{\partial\epsilon^2}\right] = \mathbb{V}\left[\frac{\partial \log f(O_i;\epsilon_0)}{\partial\epsilon}\right].$$

In fact, taking derivative over ϵ on both sides of

$$\int \frac{\partial \log f(o;\epsilon)}{\partial\epsilon} f(o;\epsilon)do = 0,$$

we have

$$\int \left[\frac{\partial^2 \log f(o;\epsilon)}{\partial\epsilon^2} f(o;\epsilon) + \frac{\partial \log f(o;\epsilon)}{\partial\epsilon}\frac{\partial f(o;\epsilon)}{\partial\epsilon}\right] do = 0,$$

which implies

$$-\int \frac{\partial^2 \log f(o;\epsilon)}{\partial\epsilon^2} f(o;\epsilon)do = \int \left[\frac{\partial \log f(o;\epsilon)}{\partial\epsilon}\right]^2 f(o;\epsilon)do.$$

Therefore, combining the above two results, we show that

$$\sqrt{n}(\widehat{\epsilon}_{\text{MLE}} - \epsilon_0) \to \mathcal{N}(0, \mathcal{I}^{-1}), \tag{7.16}$$

where $\mathcal{I} = \mathbb{V}(S_\epsilon)$ is referred to as *Fisher information.*

We can also show that the inverse of Fisher information, scaled by $1/n$, is the *Cramer-Rao bound* on the variance of any unbiased estimator of ϵ; i.e.,

$$\mathbb{V}(\widehat{\epsilon}) \geq \frac{1}{n}\mathcal{I}^{-1}, \tag{7.17}$$

for any $\widehat{\epsilon}$ such that $\mathbb{E}_{\epsilon}(\widehat{\epsilon}) = \epsilon$.

Proof of the Cramer-Rao bound (7.17):

Since $\widehat{\epsilon}$ is an unbiased estimator of ϵ, $\mathbb{E}_{\epsilon}(\widehat{\epsilon}) = \epsilon$, that is,

$$\int \widehat{\epsilon}(o_1, \ldots, o_n) \prod_{i=1}^{n} f(o_i; \epsilon) do_1 \ldots do_n = \epsilon.$$

Taking derivative over ϵ on both sides of the above equation, we have

$$\int \widehat{\epsilon}(o_1, \ldots, o_n) \frac{\partial \prod_{i=1}^{n} f(o_i; \epsilon)}{\partial \epsilon} do_1 \ldots do_n = 1,$$

which implies

$$\int \widehat{\epsilon}(o_1, \ldots, o_n) \left[\sum_{i=1}^{n} \frac{\partial \log f(o_i; \epsilon)}{\partial \epsilon}\right] \prod_{i=1}^{n} f(o_i; \epsilon) do_1 \ldots do_n = 1,$$

which is equivalent to

$$\mathbb{E}_{\epsilon}\left[\widehat{\epsilon} S_n(\epsilon)\right] = 1.$$

Then by Cauchy–Schwartz inequality, we have

$$\mathbb{E}\left[\widehat{\epsilon} S_n(\epsilon_0)\right] \leq \mathbb{V}(\widehat{\epsilon})\mathbb{V}[S_n(\epsilon_0)].$$

Noting that $\mathbb{V}[S_n(\epsilon_0)] = n\mathcal{I}$, we prove (7.17). □

From (7.16), we see the asymptotic variance of $\widehat{\epsilon}_{\mathrm{MLE}}$ is the same as the Crame-Rao bound. For this reason, we say that $\widehat{\epsilon}_{\mathrm{MLE}}$ is *efficient*. Specially,

$$\sqrt{n}(\widehat{\epsilon}_{\mathrm{MLE}} - \epsilon_0) = \frac{1}{\sqrt{n}} \sum_{i=1}^{n} \phi_{\mathrm{eff},\epsilon}(O_i) + o_p(1),$$

where $\phi_{\mathrm{eff},\epsilon}(O)$ is the *efficient influence function* in estimating ϵ_0 and

$$\phi_{\mathrm{eff},\epsilon}(O) = \mathcal{I}^{-1} S_{\epsilon}.$$

The Cramer-Rao bound is helpful for evaluating the efficiency of an estimator for estimating $\mu = \mu(\epsilon)$, where $\mu(\epsilon)$ is differentiable. Consider estimator $\widehat{\mu} = \mu(\widehat{\epsilon})$. By the delta method, we have

$$\widehat{\mu} - \mu_0 = \mu(\widehat{\epsilon}) - \mu(\epsilon_0) \doteq \frac{\partial \mu(\epsilon_0)}{\partial \epsilon}(\widehat{\epsilon} - \epsilon_0).$$

Thus, the Cramer-Rao bound in estimating μ is

$$\left[\frac{\partial\mu(\epsilon_0)}{\partial\epsilon}\right]^2\mathcal{I}^{-1}.$$

By the invariance property of MLE, $\widehat{\mu}_{\text{MLE}} = \mu(\widehat{\epsilon}_{\text{MLE}})$ is MLE of $\mu = \mu(\epsilon)$. By the delta method,

$$\sqrt{n}(\widehat{\mu}_{\text{MLE}} - \mu_0) = \frac{1}{\sqrt{n}}\sum_{i=1}^{n}\phi_{\text{eff},\mu}(O_i) + o_p(1),$$

where $\phi_{\text{eff},\mu}(O)$ is the efficient influence function in estimating μ_0 and

$$\phi_{\text{eff},\mu}(O) = \frac{\partial\mu(\epsilon_0)}{\partial\epsilon}\mathcal{I}^{-1}S_\epsilon.$$

There is an important fact about the above influence function. This fact will be used as a criterion for checking if an asymptotically linear estimator is regular soon. The fact is

$$\mathbb{E}[\phi_{\text{eff},\mu}(O)S_\epsilon] = \mathbb{E}\left[\frac{\partial\mu(\epsilon_0)}{\partial\epsilon}\mathcal{I}^{-1}S_\epsilon S_\epsilon\right] = \frac{\partial\mu(\epsilon_0)}{\partial\epsilon}\mathcal{I}^{-1}\mathbb{E}\left[S_\epsilon S_\epsilon\right] = \frac{\partial\mu(\epsilon_0)}{\partial\epsilon}. \quad (7.18)$$

Super-efficiency and regularity

We have seen the efficiency of MLE; that is, the asymptotic variance of MLE achieves the Cramer-Rao bound. Actually, there are super-efficient estimators (Newey 1990). Newey (1990) gave two examples. One is the famous example of super-efficiency that was due to Hodges. The other example is an extreme example: the estimator $\widehat{\mu} = \mu_0$, whose variance is 0, which is smaller than the Cramer-Rao bound.

These super-efficient estimators perform better than MLE only at the point of μ_0, but perform worse than MLE at any different point $\mu \neq \mu_0$. These super-efficient estimators are not useful because μ_0 is unknown to begin with. The existence of such super-efficient estimators makes it impossible for us to find the "optimal estimator" that has the smallest asymptotic variance. Therefore, one approach to this problem is to restrict our searching domain to a class of estimators satisfying certain uniformity conditions that rule out such super-efficient estimators. One useful uniformity condition is to require that the convergence of the estimator to the limiting distribution be uniform in certain shrinking neighborhoods of the true parameter value.

To define this uniformity condition precisely, consider a local data-generating process (LDGP) that generates i.i.d. observations $O_1, \ldots, O_n$ with $f(O; \epsilon_n)$, where $\sqrt{n}(\epsilon_n - \epsilon_0)$ is bounded as $n \to \infty$. As defined in Newey (1990), an estimator $\widehat{\mu}$ is said to be *regular in a parametric model* if, under any LDGP with $\{\epsilon_n\}$, $\sqrt{n}(\widehat{\mu} - \mu(\epsilon_n))$ has a limiting distribution that does not depend on sequence $\{\epsilon_n\}$.

Consider the extreme example that $\widehat{\mu} = \mu_0$. If $\mu = \epsilon$ and $\epsilon_n = \epsilon_0 + \delta/\sqrt{n}$, then $\sqrt{n}[\widehat{\mu} - \mu(\epsilon_n)] = \sqrt{n}[\epsilon_0 - (\epsilon_0 + \delta/\sqrt{n})] = -\delta$, which is a degenerate distribution that depends on the LDGP with $\{\epsilon_n\}$. Therefore, this extreme estimator is not regular.

Newey (1990) presented the **fundamental theorem of regularity**:

Theorem *Suppose that $\widehat{\mu}$ is an asymptotically linear estimator with influence function $\phi(O)$. Suppose that $\mu(\epsilon)$ is differentiable and $\mathbb{E}_\epsilon[\phi^2(O)]$ exists and is continuous on a neighborhood of ϵ_0. Then $\widehat{\mu}$ is regular if and only if*

$$\partial\mu(\epsilon_0)/\partial\epsilon^T = \mathbb{E}[\phi(O)S_\epsilon^T]. \tag{7.19}$$

By the fundamental theorem of regularity, from (7.18), we see that $\widehat{\mu}_{\text{MLE}}$ is regular. Given that we have shown that $\widehat{\mu}_{\text{MLE}}$ is asymptotically linear, we conclude that $\widehat{\mu}_{\text{MLE}}$ is a regular and asymptotically linear (RAL) estimator.

Regularity of M-estimator

Now we are ready to show that the M-estimator, which has the property of linearity (7.12), is regular, leading to the conclusion that the M-estimator is an RAL estimator. For this aim, after taking the derivative over ϵ on both sides of equation (7.7), which is equivalent to the following equation,

$$\int M(o;\mu(\epsilon))f(o;\epsilon)do = 0,$$

we obtain

$$\int \frac{\partial M(o;\mu)}{\partial\mu^{\mathrm{T}}}\frac{\partial\mu(\epsilon)}{\partial\epsilon^{\mathrm{T}}}f(o;\epsilon)do + \int M(o;\mu(\epsilon))\frac{\partial f(o;\epsilon)}{\partial\epsilon^{\mathrm{T}}}do = 0.$$

Thus, we have

$$\int \frac{\partial M(o;\mu)}{\partial\mu^{\mathrm{T}}}f(o;\epsilon)do\frac{\partial\mu(\epsilon)}{\partial\epsilon^{\mathrm{T}}} + \int M(o;\mu(\epsilon))\frac{\partial \log f(o;\epsilon)}{\partial\epsilon^{\mathrm{T}}}f(o;\epsilon)do = 0,$$

which implies

$$\int \frac{\partial M(o;\mu_0)}{\partial\mu^{\mathrm{T}}}f(o;\epsilon_0)do\frac{\partial\mu(\epsilon_0)}{\partial\epsilon^{\mathrm{T}}} + \int M(o;\mu_0)\frac{\partial \log f(o;\epsilon_0)}{\partial\epsilon^{\mathrm{T}}}f(o;\epsilon_0)do = 0,$$

which can be written as

$$\mathbb{E}\left[\frac{\partial M(O;\mu_0)}{\partial\mu^{\mathrm{T}}}\right]\frac{\partial\mu(\epsilon_0)}{\partial\epsilon^{\mathrm{T}}} + \mathbb{E}\left[M(O;\mu_0)S_\epsilon^{\mathrm{T}}\right] = 0.$$

Thus, we have

$$\frac{\partial\mu(\epsilon_0)}{\partial\epsilon^{\mathrm{T}}} = \mathbb{E}\left[-\left\{\mathbb{E}\left[\frac{\partial M(O;\mu_0)}{\partial\mu^{\mathrm{T}}}\right]\right\}^{-1}M(O;\mu_0)S_\epsilon^{\mathrm{T}}\right],$$

which is equivalent to

$$\partial\mu(\epsilon_0)/\partial\epsilon^{\mathrm{T}} = \mathbb{E}\left[\phi(O)S_\epsilon^{\mathrm{T}}\right],$$

where $\phi(O)$ is the influence function defined in (7.13). Therefore, by the fundamental theorem of regularity, we prove that the M-estimator is regular.

7.3 G-computation Estimator

7.3.1 Plug-in estimator

Consider the statistical estimand in terms of (7.2), which depends on the outcome regression function $Q(x, a)$. Thus, we can rewrite (7.2) as

$$\theta = \mathbb{E}[Q(X, 1)] - \mathbb{E}[Q(X, 0)]. \tag{7.20}$$

Hence, if we can obtain an estimator of $Q(x, a)$, denoted as $\widehat{Q}(x, a)$, and use the sample mean to estimate the population mean, then we can obtain a g-computation estimator of the statistical estimand in (7.20),

$$\widehat{\theta}_{\text{g-comp}} = \frac{1}{n}\sum_{i=1}^{n}\widehat{Q}(X_i, 1) - \frac{1}{n}\sum_{i=1}^{n}\widehat{Q}(X_i, 0), \tag{7.21}$$

where subscript "g-comp" stands for "g-computation."

In order to conduct statistical inference based on $\widehat{\theta}_{\text{g-comp}}$, we need to have an estimator of its variance. Although we can always use the bootstrap procedure (Efron and Tibshirani 1994) to estimate its variance without attempting to derive an explicit expression for it, it is desirable to derive an explicit formula for the asymptotic variance of $\widehat{\theta}_{\text{g-comp}}$, which will be derived soon.

To see that estimator $\widehat{\theta}_{\text{g-comp}}$ is a plug-in estimator, we rewrite the statistical estimand in (7.20) as

$$\theta = \int [Q(x, 1) - Q(x, 0)]dF_X(x) \triangleq \theta(Q, F_X), \tag{7.22}$$

where F_X is the probability distribution function of X.

It is well-known that the empirical distribution of $\{X_1, \ldots, X_n\}$ is the nonparametric maximum likelihood estimator (NPMLE) of F_X. The empirical distribution is defined as

$$\widehat{F}_X(x) = \frac{1}{n}\sum_{i=1}^{n} I(X_i \leq x),$$

where $X_i = (X_i(1), \ldots, X_i(p)) \le x = (x(1), \ldots, x(p))$ means $X_i(j) \le x(j)$, $j = 1, \ldots, p$. One property of the empirical distribution function is that, for any function $h(x)$, we have

$$\int h(x) d\widehat{F}_X(x) = \frac{1}{n}\sum_{i=1}^{n} h(X_i).$$

Utilizing this property, we see that

$$\begin{aligned}\widehat{\theta}_{\text{g-comp}} &= \frac{1}{n}\sum_{i=1}^{n}\left[\widehat{Q}(X_i, 1) - \widehat{Q}(X_i, 0)\right] \\ &= \int \left[\widehat{Q}(x, 1) - \widehat{Q}(x, 0)\right] d\widehat{F}_X(x) = \theta(\widehat{Q}, \widehat{F}_X).\end{aligned}$$

Therefore, $\widehat{\theta}_{\text{g-comp}} = \theta(\widehat{Q}, \widehat{F}_X)$ is a plug-in estimator of $\theta = \theta(Q, F_X)$.

7.3.2 MLE

We have seen that the g-computation estimator is a plug-in estimator. Based on this idea, we can construct any number of g-computation estimators: (1) obtain an estimator $\widehat{Q}'$ via a generalized linear model (GLM) or any other predictive model (Hastie, Tibshirani, and Friedman 2009); (2) consider empirical distribution $\widehat{F}_X$; (3) obtain a g-computation estimator $\widehat{\theta}'_{\text{g-comp}} = \theta(\widehat{Q}', \widehat{F}_X)$.

In this subsection, we focus on g-computation estimator when Q is estimated via GLM. We start with specifying a GLM, $Q(x, a; \beta)$, for $Q(x, a)$.

Continuous outcome

For continuous outcome Y, we consider the following linear model,

$$Q(X, A; \beta) = \beta(0) + \beta(1)A + \beta^{\mathrm{T}}(2)X + \beta^{\mathrm{T}}(3)AX, \tag{7.23}$$

where $\beta = (\beta(0), \beta^{\mathrm{T}}(2), \beta^{\mathrm{T}}(3))^{\mathrm{T}}$. We can obtain the minimum loss estimator (MLE) of β,

$$\widehat{\beta} = \arg\min \sum_{i=1}^{n} \left[Y_i - Q(X_i, A_i; \beta)\right]^2, \tag{7.24}$$

where the loss function is the L_2 loss function. Estimator (7.24) is equivalent to the solution of the following estimating equation,

$$\sum_{i=1}^{n} \frac{\partial Q(X_i, A_i; \beta)}{\partial \beta} \left[Y_i - Q(X_i, A_i; \beta)\right] = 0, \tag{7.25}$$

where

$$\frac{\partial Q(X,A;\beta)}{\partial \beta} = \begin{pmatrix} 1 \\ A \\ X \\ AX \end{pmatrix}. \tag{7.26}$$

Binary outcome

For binary outcome Y, we consider the following logistic model,

$$\text{logit}[Q(X,A;\beta)] = \beta(0) + \beta(1)A + \beta^{\mathrm{T}}(2)X + \beta^{\mathrm{T}}(3)AX, \tag{7.27}$$

where $\text{logit}[Q] = \log[Q/(1-Q)]$ and $\beta = (\beta(0), \beta^{\mathrm{T}}(2), \beta^{\mathrm{T}}(3))^{\mathrm{T}}$. We can obtain the maximum likelihood estimator (MLE) of β,

$$\widehat{\beta} = \arg\max \prod_{i=1}^{n} Q(X_i, 1; \beta)^{Y_i} Q(X_i, 0; \beta)^{1-Y_i}, \tag{7.28}$$

where the maximization objective function is the likelihood function. It is equivalent to the minimum loss estimator (MLE),

$$\widehat{\beta} = \arg\min \sum_{i=1}^{n} -\left[Y_i \log\{Q(X_i, 1; \beta)\} + (1-Y_i)\log\{Q(X_i, 0; \beta)\}\right], \tag{7.29}$$

where the loss function is the negative-log-likelihood loss function. Estimator (7.29) is equivalent to the solution of the following estimating equation,

$$\sum_{i=1}^{n} \frac{\partial Q(X_i, A_i; \beta)/\partial \beta}{Q(X_i, A_i; \beta)[1 - Q(X_i, A_i; \beta)]}\left[Y_i - Q(X_i, A_i; \beta)\right] = 0, \tag{7.30}$$

where

$$\frac{\partial Q(X,A;\beta)/\partial \beta}{Q(X,A;\beta)[1-Q(X,A;\beta)]} = \begin{pmatrix} 1 \\ A \\ X \\ AX \end{pmatrix}. \tag{7.31}$$

After we obtain an estimator of β, via linear model for continuous outcome or logistic model for binary outcome, we can obtain an estimator of Q, $\widehat{Q}_{\text{MLE}}(X, A) = Q(X, A; \widehat{\beta})$. Consequently, we obtain a plug-in estimator for θ,

$$\widehat{\theta}_{\text{MLE}} = \theta(\widehat{Q}_{\text{MLE}}, \widehat{F}_X) = \frac{1}{n}\sum_{i=1}^{n}\left[Q(X_i, 1; \widehat{\beta}) - Q(X_i, 0; \widehat{\beta})\right]. \tag{7.32}$$

Here is a remark on why $\widehat{\theta}_{\text{MLE}}$ is MLE of θ. This is due to the invariance property of MLE: if $h(\mu)$ is the parameter of interest and $\widehat{\mu}_{\text{MLE}}$ is MLE of μ, then $h(\widehat{\mu}_{\text{MLE}})$ is MLE of $h(\mu)$. Using the invariance property, since $\widehat{\beta}$ is MLE of β, $\widehat{Q}_{\text{MLE}}(X, A) = Q(X, A; \widehat{\beta})$ is MLE of $Q(X, A)$. Using the invariance property one more time, since $\widehat{Q}_{\text{MLE}}$ is MLE of Q and $\widehat{F}_X$ is NPMLE of F_X, $\widehat{\theta}_{\text{MLE}} = \theta(\widehat{Q}_{\text{MLE}}, \widehat{F}_X)$ is MLE of $\theta = \theta(Q, F_X)$.

7.3.3 Asymptotic variance

To derive an explicit expression of the asymptotic variance of $\widehat{\theta}_{\text{MLE}}$, we consider M-estimation. Here we consider the setting of continuous outcome, and the discussion can be extended to binary outcome easily.

Let $\mu = (\theta, \beta^{\mathrm{T}})^{\mathrm{T}}$ and $M(O; \mu) = (M_1(O, \mu), M_2^{\mathrm{T}}(O, \mu))^{\mathrm{T}}$, where

$$M_1(O, \mu) = Q(X, 1; \beta) - Q(X, 0; \beta) - \theta, \tag{7.33}$$
$$M_2(O, \mu) = [Y - Q(X, A; \beta)]\partial Q(X, A; \beta)/\partial \beta. \tag{7.34}$$

Let θ_0 be the true value of θ and β_0 be the true value of β if the parametric model $Q(X, A; \beta)$ is a correctly specified model for the outcome regression function. Then $\mu_0 = (\theta_0, \beta_0^{\mathrm{T}})^{\mathrm{T}}$ is the true value of μ.

The meat term in the sandwich formula is

$$\mathbb{E}\left\{M^{\otimes 2}(O; \mu_0)\right\} = \begin{pmatrix} \mathbb{E}[Q(X, 1; \beta_0) - Q(X, 0; \beta_0) - \theta_0]^2 & 0^{\mathrm{T}} \\ 0 & \mathbf{A}_0 \end{pmatrix}, \tag{7.35}$$

where

$$\mathbf{A}_0 = \mathbb{E}\left\{[Y - Q(X, A; \beta_0)]^2[\partial Q(X, A; \beta_0)/\partial \beta]^{\otimes 2}\right\}.$$

The bread term in the sandwich formula is

$$\mathbb{E}\left[\frac{\partial M(O, \mu_0)}{\partial \mu^{\mathrm{T}}}\right] = \begin{pmatrix} -1 & \mathbf{B}_0 \\ 0 & -\mathbf{C}_0 \end{pmatrix}, \tag{7.36}$$

whose inverse is

$$\left\{\mathbb{E}\left[\frac{\partial M(O, \mu_0)}{\partial \mu^{\mathrm{T}}}\right]\right\}^{-1} = \begin{pmatrix} -1 & -\mathbf{B}_0\mathbf{C}_0^{-1} \\ 0 & -\mathbf{C}_0^{-1} \end{pmatrix}, \tag{7.37}$$

where

$$\begin{aligned} \mathbf{B}_0 &= \mathbb{E}\left\{\partial Q(X, 1; \beta_0)/\partial \beta^{\mathrm{T}} - \partial Q(X, 0; \beta_0)/\partial \beta^{\mathrm{T}}\right\} \\ &= \mathbb{E}\left\{(1, 1, X^{\mathrm{T}}, X^{\mathrm{T}}) - (1, 0, X^{\mathrm{T}}, 0)\right\} \\ &= (0, 1, 0, \mathbb{E}\{X^{\mathrm{T}}\}) \end{aligned}$$

and

$$\mathbf{C}_0 = \mathbb{E}\{[\partial Q(X, A; \beta_0)/\partial \beta]^{\otimes 2}\}.$$

Thus, we derive the consistency and asymptotic normality of $\widehat{\theta}_{\text{MLE}}$,

$$\widehat{\theta}_{\text{MLE}} \to \theta_0, \tag{7.38}$$
$$\sqrt{N}(\widehat{\theta}_{\text{MLE}} - \theta_0) \to \mathcal{N}(0, \sigma_{0,\text{MLE}}^2), \tag{7.39}$$

where

$$\sigma_{0,\text{MLE}}^2 = \mathbb{E}[Q(X, 1; \beta_0) - Q(X, 0; \beta_0) - \theta_0]^2 + \mathbf{B}_0\mathbf{C}_0^{-1}\mathbf{A}_0\mathbf{C}_0^{-1}\mathbf{B}_0^{\mathrm{T}}, \tag{7.40}$$

which can be estimated by

$$\widehat{\sigma}^2_{\text{MLE}} = \frac{1}{n}\sum_{i=1}^{n}\left[Q(X_i, 1; \widehat{\beta}) - Q(X_i, 0; \widehat{\beta}) - \widehat{\theta}_{\text{MLE}}\right]^2 + \widehat{\mathbf{B}}\widehat{\mathbf{C}}^{-1}\widehat{\mathbf{A}}\widehat{\mathbf{C}}^{-1}\widehat{\mathbf{B}}^{\mathsf{T}},$$

where

$$\widehat{\mathbf{A}} = \frac{1}{n}\sum_{i=1}^{n}\left\{[Y_i - Q(X_i, A_i; \widehat{\beta})]^2[\partial Q(X_i, A_i; \widehat{\beta})/\partial\beta]^{\otimes 2}\right\},$$
$$\widehat{\mathbf{B}} = \frac{1}{n}\sum_{i=1}^{n}\left\{\partial Q(X_i, 1; \widehat{\beta})/\partial\beta^{\mathsf{T}} - \partial Q(X_i, 0; \widehat{\beta})/\partial\beta^{\mathsf{T}}\right\},$$
$$\widehat{\mathbf{C}} = \frac{1}{n}\sum_{i=1}^{n}[\partial Q(X_i, A_i; \widehat{\beta})/\partial\beta]^{\otimes 2}.$$

7.3.4 Influence function

To summarize, we show that $\widehat{\theta}_{\text{MLE}}$ is an RAL estimator,

$$\sqrt{n}(\widehat{\theta}_{\text{MLE}} - \theta_0) = \frac{1}{\sqrt{n}}\sum_{i=1}^{n}\phi_{\text{MLE}}(O_i) + o_p(1), \tag{7.41}$$

where $\phi_{\text{MLE}}(O)$ is the corresponding influence function and

$$\begin{aligned}\phi_{\text{MLE}}(O) =& Q(X, 1; \beta_0) - Q(X, 0; \beta_0)\\ &+ \mathbf{B}_0\mathbf{C}_0^{-1}\frac{\partial Q(X, A; \beta_0)}{\partial\beta}[Y - Q(X, A; \beta_0)] - \theta_0.\end{aligned} \tag{7.42}$$

7.4 Inverse Probability Weighted Estimator

7.4.1 IPW estimator

Consider the statistical estimand in (7.4), which depends on the propensity score function $g(a|x)$. Thus, we can rewrite (7.4) as

$$\theta = \mathbb{E}\left[\frac{I(A=1)}{g(1|X)}Y\right] - \mathbb{E}\left[\frac{I(A=0)}{g(0|X)}Y\right]. \tag{7.43}$$

Hence, if we can obtain an estimator of $g(a|x)$, denoted as $\widehat{g}(a|x)$, and use the sample mean to estimate the population mean, then we can obtain an IPW estimator of (7.43),

$$\widehat{\theta}_{\text{IPW}} = \frac{1}{n}\sum_{i=1}^{n}\frac{I(A_i=1)}{\widehat{g}(1|X_i)}Y_i - \frac{1}{n}\sum_{i=1}^{n}\frac{I(A_i=0)}{\widehat{g}(0|X_i)}Y_i. \tag{7.44}$$

7.4.2 Asymptotic variance

To derive an explicit expression of the asymptotic variance of $\widehat{\theta}_{\text{IPW}}$, we consider M-estimation. We start with specifying a parametric model for the propensity score function. For example, we may specify a logistic regression model,

$$g(1|X;\gamma) = 1 - g(0|X;\gamma) = \frac{\exp(\gamma^{\mathrm{T}}\widetilde{X})}{1+\exp(\gamma^{\mathrm{T}}\widetilde{X})}, \text{ where } \widetilde{X} = \begin{pmatrix} 1 \\ X \end{pmatrix}. \tag{7.45}$$

Based on the observed data $\{O_i = (X_i, A_i, Y_i), i = 1, \ldots, n\}$, we can obtain MLE of γ,

$$\widehat{\gamma} = \arg\max \prod_{i=1}^{n} g(1|X_i;\gamma)^{A_i} g(0|X_i;\gamma)^{1-A_i},$$

which is equivalent to

$$\widehat{\gamma} = \arg\min \sum_{i=1}^{n} - [A_i \log\{g(1|X_i;\gamma)\} + (1-A_i)\log\{g(0|X_i;\gamma)\}],$$

which is further equivalent to the solution to the following estimating equation,

$$\sum_{i=1}^{n} \frac{\partial g(1|X_i;\gamma)/\partial\gamma}{g(1|X_i;\gamma)g(0|X_i;\gamma)} [A_i - g(1|X_i;\gamma)] = 0, \tag{7.46}$$

which can be simplified as

$$\sum_{i=1}^{n} \widetilde{X}_i [A_i - g(1|X_i;\gamma)] = 0, \tag{7.47}$$

noting that

$$\frac{\partial g(1|X;\gamma)/\partial\gamma}{g(1|X;\gamma)g(0|X;\gamma)} = \widetilde{X}. \tag{7.48}$$

Since $\widehat{g}(a|X) = g(a|X,\widehat{\gamma})$, $\widehat{\theta}_{\text{IPW}}$ in (7.44) becomes

$$\widehat{\theta}_{\text{IPW}} = \frac{1}{n}\sum_{i=1}^{n} \frac{I(A_i=1)}{g(1|X_i;\widehat{\gamma})} Y_i - \frac{1}{n}\sum_{i=1}^{n} \frac{I(A_i=0)}{g(0|X_i;\widehat{\gamma})} Y_i. \tag{7.49}$$

Now we are ready to derive the asymptotic normality of $\widehat{\theta}_{\text{IPW}}$ in (7.49) within the framework of M-estimation. Let $\mu = (\theta, \gamma^{\mathrm{T}})^{\mathrm{T}}$, and $M(O;\mu) = (M_1(O,\mu), M_2^{\mathrm{T}}(O,\mu))^{\mathrm{T}}$, where

$$M_1(O,\mu) = \frac{AY}{g(1|X;\gamma)} - \frac{(1-A)Y}{g(0|X;\gamma)} - \theta, \tag{7.50}$$

$$M_2(O,\mu) = [A - g(1|X;\gamma)]\widetilde{X}. \tag{7.51}$$

Let θ_0 be the true value of θ and γ_0 be the true value of γ if the parametric model $g(a|X;\gamma)$ is a correctly specified model for the propensity score function. Then $\mu_0 = (\theta_0, \gamma_0^{\mathrm{T}})^{\mathrm{T}}$ is the true value of μ.

The meat term in the sandwich formula is

$$\mathbb{E}\left\{M^{\otimes 2}(O;\mu_0)\right\} = \begin{pmatrix} \mathbb{E}\left[\frac{AY}{g(1|X;\gamma_0)} - \frac{(1-A)Y}{g(0|X;\gamma_0)} - \theta\right]^2 & 0^{\mathrm{T}} \\ 0 & \mathbf{D}_0 \end{pmatrix},$$

where the off-diagonal elements are equal to zero by the law of iterated expectations and

$$\mathbf{D}_0 = \mathbb{E}\left\{[A - g(1|X;\gamma_0)]^2 \widetilde{X}^{\otimes 2}\right\}. \tag{7.52}$$

The bread term in the sandwich formula is

$$\mathbb{E}\left[\frac{\partial M(O,\mu_0)}{\partial \mu^{\mathrm{T}}}\right] = \begin{pmatrix} -1 & \mathbf{E}_0 \\ 0 & -\mathbf{F}_0 \end{pmatrix},$$

whose inverse is

$$\left\{\mathbb{E}\left[\frac{\partial M(O,\mu_0)}{\partial \mu^{\mathrm{T}}}\right]\right\}^{-1} = \begin{pmatrix} -1 & -\mathbf{E}_0\mathbf{F}_0^{-1} \\ 0 & -\mathbf{F}_0^{-1} \end{pmatrix}, \tag{7.53}$$

where

$$\mathbf{E}_0 = -\mathbb{E}\left\{\frac{[A - g(1|X;\gamma_0)]^2 Y}{g(1|X;\gamma_0)g(0|X;\gamma_0)}\widetilde{X}^{\mathrm{T}}\right\},$$
$$\mathbf{F}_0 = \mathbb{E}\left\{g(1|X;\gamma_0)g(0|X;\gamma_0)\widetilde{X}^{\otimes 2}\right\}.$$

Thus, if $g(a|x;\gamma)$ is a correct model for $g(a|x)$, we derive the consistency and asymptotic normality of $\widehat{\theta}_{\mathrm{IPW}}$,

$$\widehat{\theta}_{\mathrm{IPW}} \to \theta_0, \tag{7.54}$$
$$\sqrt{n}(\widehat{\theta}_{\mathrm{IPW}} - \theta_0) \to \mathcal{N}(0, \sigma_{0,\mathrm{IPW}}^2), \tag{7.55}$$

where

$$\sigma_{0,\mathrm{IPW}}^2 = \left[\frac{AY}{g(1|X;\gamma_0)} - \frac{(1-A)Y}{g(0|X;\gamma_0)} - \theta_0\right]^2 + \mathbf{E}_0\mathbf{F}_0^{-1}\mathbf{D}_0\mathbf{F}_0^{-1}\mathbf{E}_0^{\mathrm{T}}, \tag{7.56}$$

which can be estimated by

$$\widehat{\sigma}_{\mathrm{IPW}}^2 = \frac{1}{n}\sum_{i=1}^{n}\left[\frac{A_iY_i}{g(1|X_i;\widehat{\gamma})} - \frac{(1-A_i)Y_i}{g(0|X_i;\widehat{\gamma})} - \widehat{\theta}_{\mathrm{IPW}}\right]^2 + \widehat{\mathbf{E}}\widehat{\mathbf{F}}^{-1}\widehat{\mathbf{D}}\widehat{\mathbf{F}}^{-1}\widehat{\mathbf{E}}^{\mathrm{T}},$$

where

$$\widehat{\mathbf{D}} = \frac{1}{n}\sum_{i=1}^{n}[A_i - g(1|X_i;\widehat{\gamma})]^2 \widetilde{X}_i^{\otimes 2},$$

$$\widehat{\mathbf{E}} = -\frac{1}{n}\sum_{i=1}^{n}\frac{[A_i - g(1|X_i;\widehat{\gamma})]^2 Y_i}{g(1|X_i;\widehat{\gamma})g(0|X_i;\widehat{\gamma})}\widetilde{X}_i^{\mathrm{T}},$$

$$\widehat{\mathbf{F}} = \frac{1}{n}\sum_{i=1}^{n} g(1|X_i;\widehat{\gamma})g(0|X_i;\widehat{\gamma})\widetilde{X}_i^{\otimes 2}.$$

7.4.3 Influence function

To summarize, we show that $\widehat{\theta}_{\mathrm{IPW}}$ is an RAL estimator,

$$\sqrt{n}(\widehat{\theta}_{\mathrm{IPW}} - \theta_0) = \frac{1}{\sqrt{n}}\sum_{i=1}^{n}\phi_{\mathrm{IPW}}(O_i) + o_p(1), \tag{7.57}$$

where $\phi_{\mathrm{IPW}}(O)$ is the corresponding influence function and

$$\phi_{\mathrm{IPW}}(O) = \frac{AY}{g(1|X;\gamma_0)} - \frac{(1-A)Y}{g(0|X;\gamma_0)} + \mathbf{E}_0\mathbf{F}_0^{-1}\widetilde{X}[A - g(1|X;\gamma_0)] - \theta_0. \tag{7.58}$$

7.5 Augmented Inverse Probability Weighted Estimator

7.5.1 A class of estimators

We have shown that $\widehat{\theta}_{\mathrm{IPW}}$ is consistent and asymptotically normal if the propensity score function is correctly specified. Furthermore, it is desirable to construct some estimator that is not only consistent and asymptotically normal but also has the smallest asymptotic variance. We refer to such an estimator as an efficient estimator (Bickel et al. 1993; Tsiatis 2006).

Let's start with the following class of estimators,

$$\widehat{\theta}'(H) = \frac{1}{n}\sum_{i=1}^{n}\left\{\frac{A_iY_i}{g(1|X_i;\gamma_0)} - \frac{(1-A_i)Y_i}{g(0|X_i;\gamma_0)} - H(A_i,X_i)\right\},$$

where $H(A,X)$ is any function satisfying $\mathbb{E}[H(A,X)] = 0$. Because of (7.43), we can easily verify that $\widehat{\theta}(H)$ is an unbiased estimator of θ_0 for any $H(A,X)$ satisfying that $\mathbb{E}[H(A,X)] = 0$.

In addition, we can know the form of $H(A,X)$ to some extent. In fact, because A is binary and $\mathbb{E}[H(A,X)] = 0$, we have

$$H(1,X)g(1|X;\gamma_0) + H(0,X)g(0|X;\gamma_0) = 0,$$

implying that $H(1,X) = -H(0,X)g(0|X;\gamma_0)/g(1|X;\gamma_0)$. Thus, we have

$$H(1,X) = -H(0,X)[1 - g(1|X;\gamma_0)]/g(1|X;\gamma_0),$$
$$H(0,X) = -H(0,X)[0 - g(1|X;\gamma_0)]/g(1|X;\gamma_0).$$

Letting $h(X) = -H(0,X)/g(1|X;\gamma_0)$, we show that

$$H(A,X) = [A - g(1|X;\gamma_0)]h(X).$$

Therefore, class $\{\widehat{\theta}(H)\}$ is equivalent to the following class of estimators,

$$\widehat{\theta}'(h) = \frac{1}{n}\sum_{i=1}^{n}\left\{\frac{A_iY_i}{g(1|X_i;\gamma_0)} - \frac{(1-A_i)Y_i}{g(0|X_i;\gamma_0)} - [A - g(1|X;\gamma_0)]h(X_i)\right\},$$

where $h(X)$ is any function of X.

If the propensity score model is correctly specified and γ_0 is estimated by $\widehat{\gamma}$, the above class of estimators becomes the following class of estimators,

$$\widehat{\theta}(h) = \frac{1}{n}\sum_{i=1}^{n}\left\{\frac{A_iY_i}{g(1|X_i;\widehat{\gamma})} - \frac{(1-A_i)Y_i}{g(0|X_i;\widehat{\gamma})} - [A_i - g(1|X_i;\widehat{\gamma})]h(X_i)\right\}, \quad (7.59)$$

where $h(X)$ is any arbitrary function of X. Note that $\widehat{\theta}_{\text{IPW}}$ is the special case of (7.59) with $h(X) \equiv 0$.

We can easily show that for any h, $\widehat{\theta}(h)$ is consistent if the propensity score function is correctly specified. In fact, if γ_0 is the true value of γ,

$$\begin{aligned} &\mathbb{E}\left\{[A - g(1|X;\gamma_0)]h(X)\right\} \\ \overset{(a)}{=}&\mathbb{E}\left\{\mathbb{E}[A - g(1|X;\gamma_0)]h(X)|\, X\right\} \\ \overset{(b)}{=}&\mathbb{E}\left\{h(X)\mathbb{E}[A - g(1|X;\gamma_0)]|\, X\right\} = 0, \end{aligned}$$

where (a) is using the law of iterated expectations and (b) holds because $\mathbb{E}(A|X) = g(1|X;\gamma_0)$.

7.5.2 Asymptotic variances

We can apply the M-estimation theory to derive the asymptotic variance of $\widehat{\theta}(h)$. Assume logistic model (7.45) for the propensity score function. Let $\mu = (\theta, \gamma^{\mathrm{T}})^{\mathrm{T}}$ and $M_h(O;\mu) = (M_{h1}(O;\mu), M_{h2}^{\mathrm{T}}(O;\mu))^{\mathrm{T}}$, where

$$M_{h1}(O;\mu) = \frac{AY}{g(1|X;\gamma)} - \frac{(1-A)Y}{g(0|X;\gamma)} - [A - g(1|X;\gamma)]h(X) - \theta, \quad (7.60)$$

$$M_{h2}(O;\mu) = [A - g(1|X;\gamma)]\widetilde{X}. \quad (7.61)$$

The meat term in the sandwich formula is

$$\mathbb{E}\left\{M_h^{\otimes 2}(O;\mu_0)\right\} = \begin{pmatrix} V_0(h) & 0^{\mathrm{T}} \\ 0 & \mathbf{D}_0 \end{pmatrix},$$

where $\mathbf{D}_0$ is the same as the one defined in the preceding subsection and

$$
\begin{aligned}
V_0(h) =& \mathbb{E}\left[\frac{AY}{g(1|X;\gamma_0)} - \frac{(1-A)Y}{g(0|X;\gamma_0)} - \{A - g(1|X;\gamma_0)\}h(X) - \theta_0\right]^2 \\
=& \mathbb{E}\left[\frac{AY}{g(1|X;\gamma_0)} - \frac{(1-A)Y}{g(0|X;\gamma_0)} - \{A - g(1|X;\gamma_0)\}h(X)\right]^2 - \theta_0^2 \\
=& \mathbb{E}\left[\frac{AY}{g(1|X;\gamma_0)} - \frac{(1-A)Y}{g(0|X;\gamma_0)}\right]^2 + \mathbb{E}\left[\{A - g(1|X;\gamma_0)\}h(X)\right]^2 \\
-& 2\mathbb{E}\left[\left\{\frac{AY}{g(1|X;\gamma_0)} - \frac{(1-A)Y}{g(0|X;\gamma_0)}\right\}\{A - g(1|X;\gamma_0)\}h(X)\right] - \theta_0^2. \quad (7.62)
\end{aligned}
$$

The bread term in the sandwich formula is

$$
\mathbb{E}\left[\frac{\partial M(O,\mu_0)}{\partial \mu^{\mathrm{T}}}\right] = \begin{pmatrix} -1 & \mathbf{E}_0(h) \\ 0 & -\mathbf{F}_0 \end{pmatrix},
$$

where $\mathbf{F}_0$ is the same as the one defined in the preceding subsection and

$$
\begin{aligned}
\mathbf{E}_0(h) =& \mathbb{E}\Bigg\{\Bigg[-\frac{AYg(0|X_0;\gamma_0)}{g(1|X;\gamma_0)} - \frac{(1-A)Yg(1|X;\gamma_0)}{g(0|X;\gamma_0)} \\
& + g(1|X;\gamma_0)g(0|X;\gamma_0)h(X)\Bigg]\widetilde{X}^{\mathrm{T}}\Bigg\} \\
=& \mathbb{E}\Bigg\{\Bigg[-\mathbb{E}(Y|X, A=1)g(0|X_0;\gamma_0) - \mathbb{E}(Y|X, A=0)g(1|X;\gamma_0) \\
& + g(1|X;\gamma_0)g(0|X;\gamma_0)h(X)\Bigg]\widetilde{X}^{\mathrm{T}}\Bigg\}.
\end{aligned}
$$

Thus, if $g(a|x;\gamma)$ is a correct model for $g(a|x)$, we derive the asymptotic normality of $\widehat{\theta}(h)$ for any given h,

$$
\sqrt{n}(\widehat{\theta}(h) - \theta_0) \to \mathcal{N}(0, \sigma_0^2(h)), \quad (7.63)
$$

where

$$
\sigma_0^2(h) = V_0(h) + \mathbf{E}_0(h)\mathbf{F}_0^{-1}\mathbf{D}_0\mathbf{F}_0^{-1}\mathbf{E}_0^{\mathrm{T}}(h). \quad (7.64)
$$

7.5.3 AIPW estimator

Our goal is to find an optimal h^* that minimizes $\sigma_0^2(h)$. Note that the second term on the right-hand-side (RHS) of (7.64) is non-negative. Therefore, if we could find an h^* that minimizes $V_0(h)$ and satisfies $\mathbf{E}_0(h^*) = 0$, then this h^* minimizes $\sigma_0^2(h)$.

In the expansion of $V_0(h)$ in (7.62), there are four terms in the foremost RHS of (7.62), among which only two terms depend on $h(X)$. Thus, applying the law of iterated expectations to those two terms respectively, we have

$$
\begin{aligned}
&\mathbb{E}\left\{[A - g(1|X;\gamma_0)]h(X)\right\}^2 \\
=&\mathbb{E}\left(\mathbb{E}\left\{[A - g(1|X;\gamma_0)]h(X)\right\}^2 |X\right) \\
=&\mathbb{E}\left(h^2(X)\mathbb{E}\{[A - g(1|X;\gamma_0)]^2|X\}\right) \\
=&\mathbb{E}\left[g(1|X;\gamma_0)g(0|X;\gamma_0)h^2(X)\right]
\end{aligned}
\tag{7.65}
$$

and

$$
\begin{aligned}
&\mathbb{E}\left[\left\{\frac{AY}{g(1|X;\gamma_0)} - \frac{(1-A)Y}{g(0|X;\gamma_0)}\right\}\{A - g(1|X;\gamma_0)\}h(X)\right] \\
=&\mathbb{E}\left(\mathbb{E}\left[\left\{\frac{AY}{g(1|X;\gamma_0)} - \frac{(1-A)Y}{g(0|X;\gamma_0)}\right\}\{A - g(1|X;\gamma_0)\}h(X)\Big|X\right]\right) \\
=&\mathbb{E}\left(\mathbb{E}\left[\left\{\frac{Y}{g(1|X;\gamma_0)}\right\}g(0|X;\gamma_0)\Big|X, A=1\right]\mathbb{P}(A=1|X)h(X)\right) \\
&+\mathbb{E}\left(\mathbb{E}\left[\left\{\frac{Y}{g(0|X;\gamma_0)}\right\}g(1|X;\gamma_0)\Big|X, A=0\right]\mathbb{P}(A=0|X)h(X)\right) \\
=&\mathbb{E}\left\{[\mathbb{E}(Y|X, A=1)g(0|X;\lambda_0) + \mathbb{E}(Y|X, A=0)g(1|X;\lambda_0)]h(X)\right\}.
\end{aligned}
\tag{7.66}
$$

Combining the above two results, (7.65) and (7.66), we see that the optimal $h^*(X)$ that minimizes $V_0(h)$ is equal to

$$
\begin{aligned}
&h^*(X) = \arg\min_{h(\cdot)} V_0(h) = \arg\min_{h(\cdot)} \mathbb{E}\Big\{g(1|X;\gamma_0)g(0|X;\gamma_0)h^2(X) \\
&-2\big[\mathbb{E}(Y|X, A=1)g(0|X;\lambda_0) + \mathbb{E}(Y|X, A=0)g(1|X;\lambda_0)\big]h(X)\Big\}.
\end{aligned}
\tag{7.67}
$$

Note the solution of the minimization of quadratic function $f(x) = ax^2 + bx + c$ is $x^* = -b/(2a)$, as illustrated in Figure 7.1. Thus, we obtain

$$
h^*(X) = \frac{\mathbb{E}(Y|X, A=1)}{g(1|X;\gamma_0)} + \frac{\mathbb{E}(Y|X, A=0)}{g(0|X;\gamma_0)}.
\tag{7.68}
$$

In addition, we can easily verify that $h^*(X)$ satisfies $\mathbf{E}_0(h^*) = 0$. Therefore, we complete the proof that $h^*(X)$ defined in (7.68) minimizes $\sigma_0^2(h)$. Thus, the optimal "estimator" of the form (7.59) with the smallest variance is

$$
\begin{aligned}
\widehat{\theta}(h^*) = \frac{1}{n}\sum_{i=1}^{n}\Bigg\{&\frac{A_iY_i}{g(1|X_i;\widehat{\gamma})} - \frac{(1-A_i)Y_i}{g(0|X_i;\widehat{\gamma})} - [A_i - g(1|X_i;\widehat{\gamma})] \\
&\times\left(\frac{Q(X_i,1)}{g(1|X_i,\gamma_0)} + \frac{Q(X_i,0)}{g(0|X_i;\gamma_0)}\right)\Bigg\}.
\end{aligned}
\tag{7.69}
$$

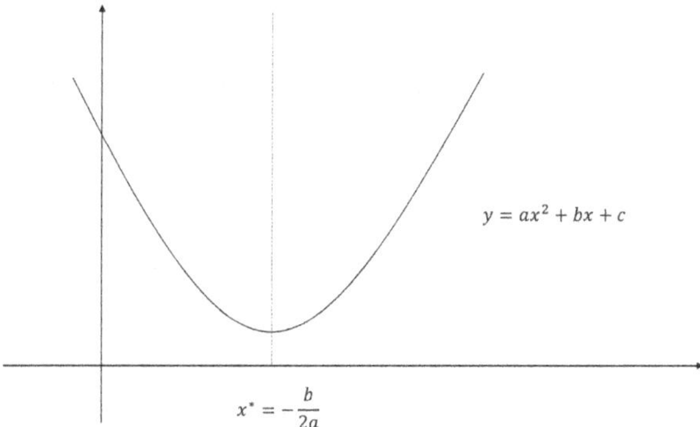

FIGURE 7.1
The axis of symmetry

This is not an estimator because it depends on the unknown outcome regression function $Q(x, a)$. If we specify a parametric model, say $Q(X, A; \beta)$, and estimate β using $\widehat{\beta}$, we construct the augmented inverse probability weighted estimator (AIPW),

$$\widehat{\theta}_{\text{AIPW}} = \frac{1}{n}\sum_{i=1}^{n}\left\{\frac{A_iY_i}{g(1|X_i;\widehat{\gamma})} - \frac{(1-A_i)Y_i}{g(0|X_i;\widehat{\gamma})} - [A_i - g(1|X_i;\widehat{\gamma})] \times \left(\frac{Q(X_i,1;\widehat{\beta})}{g(1|X_i,\widehat{\gamma})} + \frac{Q(X_i,0;\widehat{\beta})}{g(0|X_i,\widehat{\gamma})}\right)\right\}. \tag{7.70}$$

7.5.4 Double robustness

Let's put the construction of $\widehat{\theta}_{\text{AIPW}}$ within the framework of M-estimation. Assume the logistic model (7.45) for propensity score function $g(A|X)$; that is, $\text{logit}[g(A|X;\beta)] = \gamma(0) + \gamma^{\text{T}}(1)X = \gamma^{\text{T}}\widetilde{X}$, where $\gamma = (\gamma(0), \gamma^{\text{T}}(1))^{\text{T}}$.

In addition, assume a model for the outcome regression function, say linear model $Q(X, A; \beta) = \beta(0) + \beta(1)A + \beta^{\text{T}}(2)X + \beta^{\text{T}}(3)AX$ for continuous outcome and logistic model $\text{logit}[Q(X, A; \beta)] = \beta(0) + \beta(1)A + \beta^{\text{T}}(2)X + \beta^{\text{T}}(3)AX$ for binary outcome, respectively, where $\beta = (\beta(0), \beta^{\text{T}}(2), \beta^{\text{T}}(3))^{\text{T}}$. Hereafter, we consider the setting of continuous outcome only.

Let $\mu = (\theta, \beta^{\text{T}}, \gamma^{\text{T}})^{\text{T}}$ and $M(O; \mu) = (M_1(O; \mu), M_2^{\text{T}}(O; \mu), M_3^{\text{T}}(O; \mu))^{\text{T}}$ with

$$M_1(O;\mu) = \frac{AY}{g(1|X;\gamma)} - \frac{(1-A)Y}{g(0|X;\gamma)} - \{A - g(1|X;\gamma)\} \times \left[\frac{Q(X,1;\beta)}{g(1|X;\gamma)} + \frac{Q(X,0;\beta)}{g(0|X;\gamma)}\right] - \theta, \tag{7.71}$$

$$M_2(O;\mu) = \frac{\partial Q(X,A;\beta)}{\partial \beta^{\text{T}}}[Y - Q(X,A;\beta)], \tag{7.72}$$

$$M_3(O;\mu) = \widetilde{X}[A - g(1|X;\gamma)]. \tag{7.73}$$

If the propensity score model g is correctly specified, by the M-estimation theory, we have $\widehat{\gamma} \to \gamma_0$. Then, we can verify that

$$\mathbb{E}\left[M_1(O; \theta_0, \beta, \gamma_0)\right] = 0, \tag{7.74}$$

for any β. Thus, $\widehat{\theta}_{\text{AIPW}}$ is consistent if g is correctly specified.

If the outcome regression model Q is correctly specified, by the M-estimation theory, $\widehat{\beta} \to \beta_0$. Then, we can verify that

$$\mathbb{E}\left[M_1(O; \theta_0, \beta_0, \gamma)\right] = 0, \tag{7.75}$$

for any γ. Thus, $\widehat{\theta}_{\text{AIPW}}$ is consistent if Q is correctly specified.

The above property is called *double robustness* because $\widehat{\theta}_{\text{AIPW}}$ is consistent if either Q or g is correctly specified.

When both Q and g are correctly specified, by the M-estimation theory, we can show that (the details are omitted; see Ex 7.4)

$$\sqrt{n}(\widehat{\theta}_{\text{AIPW}} - \theta_0) \to \mathcal{N}(0, \sigma^2_{0,\text{AIPW}}), \tag{7.76}$$

where

$$\begin{aligned}\sigma^2_{0,\text{AIPW}} =& \mathbb{E}\left\{\frac{AY}{g(1|X;\gamma_0)} - \frac{(1-A)Y}{g(0|X;\gamma_0)} - [A - g(1|X;\gamma_0)]\right. \\ & \left. \times \left[\frac{Q(X,1;\beta_0)}{g(1|X;\gamma_0)} + \frac{Q(X,0;\beta_0)}{g(0|X;\gamma_0)}\right] - \theta_0\right\}^2,\end{aligned} \tag{7.77}$$

which can be estimated by

$$\begin{aligned}\widehat{\sigma}^2_{\text{AIPW}} =& \frac{1}{n}\sum_{i=1}^{n}\left\{\frac{A_iY_i}{g(1|X_i;\widehat{\gamma})} - \frac{(1-A_i)Y_i}{g(0|X_i;\widehat{\gamma})} - [A_i - g(1|X_i;\widehat{\gamma})]\right. \\ & \left. \times \left[\frac{Q(X_i,1;\widehat{\beta})}{g(1|X_i;\widehat{\gamma})} + \frac{Q(X_i,0;\widehat{\beta})}{g(0|X_i;\widehat{\gamma})}\right] - \widehat{\theta}_{\text{AIPW}}\right\}^2.\end{aligned}$$

7.5.5 Influence function

To summarize, we show that $\widehat{\theta}_{\text{AIPW}}$ is an RAL estimator,

$$\sqrt{n}(\widehat{\theta}_{\text{AIPW}} - \theta_0) = \frac{1}{\sqrt{n}}\sum_{i=1}^{n}\phi_{\text{AIPW}}(O_i) + o_p(1), \tag{7.78}$$

where $\phi_{\text{AIPW}}(O)$ is the corresponding influence function and

$$\begin{aligned}\phi_{\text{AIPW}}(O) =& \frac{AY}{g(1|X;\gamma_0)} - \frac{(1-A)Y}{g(0|X;\gamma_0)} - [A - g(1|X;\gamma_0)] \\ & \times \left[\frac{Q(X,1;\beta_0)}{g(1|X;\gamma_0)} + \frac{Q(X,0;\beta_0)}{g(0|X;\gamma_0)}\right] - \theta_0.\end{aligned} \tag{7.79}$$

TABLE 7.2
A small dataset

ID	X	A	Y	ID	X	A	Y
1	0	0	0	11	1	0	1
2	0	0	1	12	1	0	0
3	0	0	1	13	1	1	0
4	0	0	0	14	1	1	1
5	0	1	1	15	1	1	1
6	0	1	1	16	1	1	0
7	0	1	0	17	1	1	0
8	0	1	1	18	1	1	0
9	1	0	0	19	1	1	1
10	1	0	0	20	1	1	0

Since $\widehat{\theta}_{\text{AIPW}}$ is an efficient estimator if both g and Q are correctly specified, we call the above influence function, $\phi_{\text{AIPW}}(O)$, the efficient influence function (EIF). In the next chapter, we will see that this EIF plays a crucial role in the targeted learning framework.

In this chapter, we use only elementary calculus to derive the EIF (7.79) without stepping into the territory of the semiparametric theory. Please refer to Bickel et al. (1993) or Tsiatis (2006) for a much more elegant theory of deriving the EIF (7.79).

7.6 Exercises

Ex 7.1

Consider the small dataset in Table 7.2. Since all the three variables are binary, we can fit a saturate model for Q. Obtain a point estimate for the ATE estimand using the g-computation estimator $\widehat{\theta}_{\text{g-comp}}$.

Ex 7.2

Consider the small dataset in Table 7.2. Since all the three variables are binary, we can fit a saturate model for g. Obtain a point estimate for the ATE estimand using the IPW estimator $\widehat{\theta}_{\text{IPW}}$.

Ex 7.3

Consider the small dataset in Table 7.2. Since all the three variables are binary, we can fit saturate models for Q and g, respectively. Obtain a point estimate for the ATE estimand using the AIPW estimator $\hat{\theta}_{\text{AIPW}}$.

Ex 7.4

Following the M-estimation theory and (7.71)–(7.73), verify the result (7.76).

8

The Art of Estimation (II): TMLE

8.1 Semiparametric Statistics

8.1.1 Semiparametric estimators

Continue the discussion in Chapter 7. Consider a cohort study, in which we are interested in estimating the average treatment effect (ATE),

$$\theta^* = \mathbb{E}(Y^{a=1} - Y^{a=0}), \tag{8.1}$$

and we have data $\{O_i = (X_i, A_i, Y_i), i = 1, \ldots, n\}$.

Recall that, under the identifiability assumptions—consistency, exchangeability, and positivity—we can translate the causal estimand θ^* to a statistical estimand θ and there are two strategies to do so: the standardization strategy and the weighting strategy. By the standardization strategy, we derive the following statistical estimand,

$$\theta = \mathbb{E}[Q(X, 1) - Q(X, 0)], \tag{8.2}$$

where $Q(x, a) = \mathbb{E}(Y|X = x, A = a)$ is the outcome regression function. By the weighting strategy, we derive the following statistical etsimand,

$$\theta = \mathbb{E}\left[\frac{(2A-1)Y}{g(A|X)}\right], \tag{8.3}$$

where $g(a|x) = \mathbb{P}(A = a|X = x)]$ is the propensity score function.

In Chapter 7, within the framework of M-estimation, we reviewed three methods: MLE, IPW, and AIPW. In the construction of the MLE, we assume a parametric model $Q(x, a; \beta)$ for $Q(x, a)$; if $Q(x, a; \beta)$ is a correct model for $Q(x, a)$, then $\widehat{\theta}_{\text{MLE}}$ is a consistent estimator for θ. In the construction of the IPW estimator, we assume a parametric model $g(a|x; \gamma)$ for $g(a|x)$; if $g(a|x; \gamma)$ is a correct model for $g(a|x; \gamma)$, then $\widehat{\theta}_{\text{IPW}}$ is a consistent estimator for θ. In the construction of the AIPW estimator, we assume a parametric model $Q(x, a; \beta)$ for $Q(x, a)$ and a parametric model $g(a|x; \gamma)$ for $g(a|x)$; if either $Q(x, a; \beta)$ or $g(a|x; \gamma)$ is correct, then $\widehat{\theta}_{\text{AIPW}}$ is a consistent estimator for θ.

DOI: 10.1201/9781003433378-8

In practice, since $Q(x,a)$ and $g(a|x)$ are unknown functions, it is unrealistic to specify a parametric model for either $Q(x,a)$ or $g(a|x)$. Actually, we don't even know which variables should be included in X in the first place. This is the problem of misspecification. Therefore, to solve the problem of misspecification, it is more realistic to consider nonparametric or semiparametric models for $Q(x,a)$ and $g(a|x)$.

Let $\widehat{Q}(x,a)$ and $\widehat{g}(a|x)$ be some estimators of $Q(x,a)$ and $g(a|x)$, respectively, via some nonparametric or semiparametric predictive modeling procedures (Hastie, Tibshirani, and Friedman 2009; van der Laan, Polley, and Hubbard 2007). Based on $\widehat{Q}(x,a)$ and/or $\widehat{g}(a|x)$, we can construct the corresponding MLE, IPW estimator, and AIPW estimator:

$$
\begin{aligned}
\widehat{\theta}_{\mathrm{MLE}} &= \frac{1}{n}\sum_{i=1}^{n}\left[\widehat{Q}(X_i,1)-\widehat{Q}(X_i,0)\right],\\
\widehat{\theta}_{\mathrm{IPW}} &= \frac{1}{n}\sum_{i=1}^{n}\frac{(2A_i-1)Y_i}{\widehat{g}(A_i|X_i)},\\
\widehat{\theta}_{\mathrm{AIPW}} &= \frac{1}{n}\sum_{i=1}^{n}\left[\frac{(2A_i-1)Y_i}{\widehat{g}(1|X_i)}-\{A_i-\widehat{g}(1|X_i)\}\sum_{j=0}^{1}\frac{\widehat{Q}(X_i,j)}{\widehat{g}(j|X_i)}\right].
\end{aligned}
$$

In semiparametric statistics, a semiparametric estimator $\widehat{\theta}$ for θ is one that is consistent and asymptotically normal:

$$
\begin{aligned}
\widehat{\theta} &\xrightarrow{p} \theta,\\
\sqrt{n}(\widehat{\theta}-\theta_0) &\xrightarrow{d} \mathcal{N}(0,\sigma^2).
\end{aligned}
$$

We will demonstrate that these estimators, $\widehat{\theta}_{\mathrm{MLE}}$, $\widehat{\theta}_{\mathrm{IPW}}$, and $\widehat{\theta}_{\mathrm{AIPW}}$, are semiparametric estimators if $\widehat{Q}(x,a)$ and/or $\widehat{g}(a|x)$ are consistent. We will also demonstrate there is still room to improve their performance, which will be improved by the targeted learning approach (van der Laan and Rose 2011).

8.1.2 Super learner

There are a variety of methods for obtaining nonparametric or semiparametric estimators for $Q(x,a)$ and $g(a|x)$ (Hastie, Tibshirani, and Friedman 2009). Among them, super learner is a promising method (van der Laan, Polley, and Hubbard 2007).

Unlike parametric modeling that relies on a pre-specified parametric model, super learner utilizes a library of predictive models—statistical learning algorithms or machine learning algorithms (Hastie, Tibshirani, and Friedman 2009). Super learner attempts to choose the "best" algorithm that achieves the "best" performance in terms of a loss function. For example, for continuous outcome, we consider the L_2 loss function,

$$
L(O,Q) = [Y - Q(X,A)]^2, \tag{8.4}
$$

and for binary outcome, we consider the negative-log-likelihood loss function,

$$L(O, Q) = -\log\left\{Q(X, A)^Y [1 - Q(X, A)]^{1-Y}\right\}. \tag{8.5}$$

For treatment variable A, which is a binary variable, we also consider the negative-log-likelihood loss function,

$$L(O, g) = -\log\left\{g(1|X)^A [1 - g(0|X)]^{1-A}\right\}. \tag{8.6}$$

Super learner uses cross-validation to choose the "best" algorithm. Without cross-validation, we would choose an algorithm that overfits. For example, for continuous outcome, we may choose an algorithm that outputs $\widehat{Q}(X_i, A_i) = Y_i$ for any i, $i = 1, \ldots, n$, but such an algorithm is useless. With cross-validation, we will choose an algorithm that achieves the "best" compromise between goodness-of-fit and model complexity.

The K-fold cross-validation procedure divides the data into K subgroups, with each subgroup having about n/K observations. Leaving out one of the K subgroups as the validation set, denoted as $\mathcal{O}^{(k)}$, we fit each algorithm indicated by j in the library consisting of J algorithms to the remaining $K - 1$ subgroups, providing a fitted model denoted as $\widehat{Q}^{(j,k)}$, $k = 1, \ldots, K$ and $j = 1, \ldots, J$. For algorithm j, the cross-validated risk is estimated by

$$CV^{(j)} = \frac{1}{K} \sum_{k=1}^{K} \frac{1}{\#(\mathcal{O}^{(k)})} \sum_{O \in \mathcal{O}^{(k)}} L(O, \widehat{Q}^{(j,k)}). \tag{8.7}$$

Discrete super learner (van der Laan, Polley, and Hubbard 2007) chooses the algorithm with the smallest cross-validated risk,

$$j^* = \arg \min_{j=1,\ldots,J} CV^{(j)}.$$

van der Laan, Polley, and Hubbard (2007) further proposed super learner to improve discrete super learner. Super learner searches for an optimally weighted combination of J algorithms in the library, by minimizing the following weighted cross-validated risk,

$$CV(w_1, \ldots, w_J) = \frac{1}{K} \sum_{k=1}^{K} \frac{1}{\#(\mathcal{O}^{(k)})} \sum_{O \in \mathcal{O}^{(k)}} L(O, \sum_{j=1}^{J} w_j \widehat{Q}^{(j,k)}), \tag{8.8}$$

where $w_j \geq 0$ and $\sum_{j=1}^{J} w_j = 1$.

Both discrete super learner and super learner can be implemented using R package "SuperLearner."

8.1.3 Semiparametric estimators based on super learner

Let $\widehat{Q}_{\text{SL}}(x, a)$ and $\widehat{g}_{\text{SL}}(a|x)$ be estimators of $Q(x, a)$ and $g(a|x)$, respectively, via super learner. Based on $\widehat{Q}_{\text{SL}}(x, a)$ and/or $\widehat{g}_{\text{SL}}(a|x)$, we can construct the corresponding MLE-SL, IPW-SL, and AIPW-SL estimators:

$$\widehat{\theta}_{\text{MLE-SL}} = \frac{1}{n}\sum_{i=1}^{n}\left[\widehat{Q}_{\text{SL}}(X_i,1) - \widehat{Q}_{\text{SL}}(X_i,0)\right], \tag{8.9}$$

$$\widehat{\theta}_{\text{IPW-SL}} = \frac{1}{n}\sum_{i=1}^{n}\left[\frac{(2A_i-1)Y_i}{\widehat{g}_{\text{SL}}(A_i|X_i)}\right], \tag{8.10}$$

$$\widehat{\theta}_{\text{AIPW-SL}} = \frac{1}{n}\sum_{i=1}^{n}\left[\frac{(2A_i-1)Y_i}{\widehat{g}_{\text{SL}}(A_i|X_i)} - \{A_i - \widehat{g}_{\text{SL}}(1|X_i)\}\sum_{j=0}^{1}\frac{\widehat{Q}_{\text{SL}}(X_i,j)}{\widehat{g}_{\text{SL}}(j|X_i)}\right]. \tag{8.11}$$

Despite that we can use the bootstrap method—which is time-consuming—to obtain the variances of these estimators, it is desirable to derive explicit formulae for their asymptotic variances, which will be derived in the next section.

8.2 Asymptotic Variances of Semiparametric Estimators

"Knowledge of the asymptotic variance of an estimator is important for large sample inference, efficiency, and as a guide to the specification of regularity conditions."—Newey (1994)

8.2.1 Parametric submodels

In Chapter 7, we studied the asymptotic variances of parametric M-estimators. To study the asymptotic variances of semiparametric M-estimators, we consider the idea of *parametric submodels* (Stein 1956).

"One could imagine that the data are generated by a parametric model that satisfies the semiparametric assumptions and contains the truth. Such a model is referred to as a *parametric submodel*, where the 'sub' prefix refers to the fact that it is a subset of the model consisting of all distributions satisfying the assumptions."—Newey (1990)

Assume the true model for X is $f_{X0}(x)$, the true model for $A|(X = x)$ is $g_0(a|x)$, and the true model for $Y|(X = x, A = a)$ is $f_{Y0}(y|a, x)$. These distributions are unknown and we don't put any constraints on them. Note that the true outcome regression function is $Q_0(x, a) = \int y f_{Y0}(y|x, a)dy$.

Imagine any parametric submodel: $X \sim f_X(x; \epsilon_1)$, $A|(X = x) \sim g(a|x; \epsilon_2)$, and $Y|(X = x, A = a) \sim f_Y(y|x, a; \epsilon_3)$, such that $f_X(x; 0) = f_{X0}(x)$,

$g(a|x;0) = g_0(a|x)$, and $f_Y(y|a,x;0) = f_{Y0}(y|a,x)$. Note that the outcome regression function is then $Q(x,a;\epsilon_3) = \int y f_Y(y|x,a;\epsilon_3)dy$ and $Q(x,a;0) = Q_0(x,a)$. To be more accurate, we only imagine some *regular parametric model*, the one that is smooth with a non-singular information matrix. But we omit such technical detail here, which can be found in Newey (1990).

Under a given parametric submodel, we can calculate the corresponding score $S_{\boldsymbol{\epsilon}} = (S_{\epsilon_1}, S_{\epsilon_2}, S_{\epsilon_3})^{\mathrm{T}}$, where $\boldsymbol{\epsilon} = (\epsilon_1, \epsilon_2, \epsilon_3)^{\mathrm{T}}$ and

$$\begin{aligned}
S_{\epsilon_1} &= \partial \log f_X(x;\epsilon_1)/\partial\epsilon_1|_{\epsilon_1=0},\\
S_{\epsilon_2} &= \partial \log g(a|x;\epsilon_2)/\partial\epsilon_2|_{\epsilon_2=0},\\
S_{\epsilon_3} &= \partial \log f_Y(y|x,a;\epsilon_3)/\partial\epsilon_3|_{\epsilon_3=0}.
\end{aligned}$$

Because the expectation of any score is zero, S_{ϵ_1} is a function of X with $\mathbb{E}(S_{\epsilon_1}) = 0$, S_{ϵ_2} is a function of (X, A) with $\mathbb{E}(S_{\epsilon_2}|X) = 0$, and S_{ϵ_3} is a function of (X, A, Y) with $\mathbb{E}(S_{\epsilon_3}|X, A) = 0$. Because we could imagine any number of parametric submodels, we put all the possible S_{ϵ_j}'s into a set named as $\mathcal{T}_j$, $j = 1, 2, 3$. Thus, we show that

$$\begin{aligned}
\mathcal{T}_1 &= \{\alpha_1(X) : \mathbb{E}[\alpha_1(X)] = 0\},\\
\mathcal{T}_2 &= \{\alpha_2(X,A) : \mathbb{E}[\alpha_2(X,A)|X] = 0\},\\
\mathcal{T}_3 &= \{\alpha_3(X,A,Y) : \mathbb{E}[\alpha_3(X,A,Y)|X,A] = 0\}.
\end{aligned}$$

Here is a remark on $\mathcal{T}_2$. For any $\alpha_2(X, A)$ such that $\mathbb{E}[\alpha_2(X, A)|X] = 0$, because A is binary, we have $\alpha_2(X,1)g_0(1|X) + \alpha_2(X,0)g_0(0|X) = 0$, implying that $\alpha_2(X,1) = [-\alpha_2(X,0)/g_0(1|X)](1 - g_0(1|X)$ and $\alpha_2(X,0) = [-\alpha_2(X,0)/g_0(0|X)](-g_0(1|X)$. Then, $\alpha_2(X, A)$ can be written as $a_2(X)[A - g_0(1|X)]$, where $a_2(X) = -\alpha_2(X,0)/g_0(1|X)$. Thus, we have

$$\mathcal{T}_2 = \{[A - g_0(1|X)]a_2(X) : \text{any } a_2(X)\}.$$

We are interested in estimating the following statistical estimand:

$$\begin{aligned}
\theta_0 &= \mathbb{E}\left[Q_0(X,1) - Q_0(X,0)\right]\\
&= \int \left[\int y f_{Y0}(y|1,x)dy - \int y f_{Y0}(y|0,x)dy\right] f_{X0}(x)dx.
\end{aligned}$$

or equivalently,

$$\begin{aligned}
\theta_0 &= \mathbb{E}\left[\frac{(2A-1)Y}{g_0(A|X)}\right]\\
&= \int \left[\int y f_{Y0}(y|1,x)dy - \int y f_{Y0}(y|0,x)dy\right] f_{X0}(x)dx.
\end{aligned}$$

For any given regular parametric submodel indicated by $\boldsymbol{\epsilon} = (\epsilon_1, \epsilon_2, \epsilon_3)^{\mathrm{T}}$, the estimand associated with the parametric submodel is

$$\begin{aligned}\theta(\boldsymbol{\epsilon}) &= \mathbb{E}_{\boldsymbol{\epsilon}}\left[Q(Y|X,1;\epsilon_3) - Q(Y|X,0;\epsilon_3)\right] = \mathbb{E}_{\boldsymbol{\epsilon}}\left[\frac{(2A-1)Y}{g(A|X;\epsilon_2)}\right] \\ &= \int \left[\int y f_Y(y|x,1;\epsilon_3)dy - \int y f_Y(y|x,0;\epsilon_3)dy\right] f_X(x;\epsilon_1)dx.\end{aligned}$$

To summarize, we consider a semiparametric model where θ is the parameter of interest and there are no constraints on f_{X0}, g_0, and f_{Y0}. For any imagined parametric submodel, the estimand becomes $\theta = \theta(\boldsymbol{\epsilon})$, where $\boldsymbol{\epsilon} = (\epsilon_1, \epsilon_2, \epsilon_3)^{\mathrm{T}}$. When $\boldsymbol{\epsilon} = \mathbf{0} = (\mathbf{0}, \mathbf{0}, \mathbf{0})^{\mathrm{T}}$, $\theta_0 = \theta(\mathbf{0})$.

We use the idea of parametric submodels to define the Cramer-Rao bound for a semiparametric model. For any parametric submodel, we can obtain the Cramer-Rao bound following the discussion in Chapter 7. We aim for constructing semiparametric estimators whose asymptotic variances are comparable to the Cramer-Rao bound of the semiparametric model, which is to be defined. For any given parametric submodel, the asymptotic variances of these semiparametric estimators are no smaller than the Cramer-Rao bound of the given parametric submodel. Therefore, the asymptotic variances of these semiparametric estimators are no smaller than the supremum of the Cramer-Rao bounds of all parametric submodels—this supremum is defined as the Cramer-Rao bound of the semiparametric model.

8.2.2 The fundamental theorem of regularity

We also use the idea of parametric submodels to define the regularity. An estimator is said to be *regular* if it is regular in every regular parametric submodel and the limiting distribution does not depend on the parametric submodel (Newey 1990).

Newey (1990) presented the following fundamental theorem of regularity:

Theorem *Suppose that $\widehat{\theta}$ is an asymptotically linear estimator for θ with influence function $\phi(O)$. Suppose that, for any regular parametric submodel, $\theta(\boldsymbol{\epsilon})$ is differentiable and $\mathbb{E}_{\boldsymbol{\epsilon}}[\phi^2(O)]$ exists and is continuous on a neighborhood of $\mathbf{0}$. Then $\widehat{\theta}$ is regular if and only if, for any regular parametric submodel,*

$$\partial\theta(\boldsymbol{\epsilon})/\partial\boldsymbol{\epsilon}\Big|_{\boldsymbol{\epsilon}=\mathbf{0}} = \mathbb{E}[\phi(O)S_{\boldsymbol{\epsilon}}], \tag{8.12}$$

where $S_{\boldsymbol{\epsilon}}$ is the score of the corresponding parametric submodel.

The first direct application of the fundamental theorem of regularity

Let $U_1, \ldots, U_n$ be observations which are i.i.d. with $U \sim f_0(u)$. The estimand of interest is $\beta_0 = \mathbb{E}(U)$. Let the distribution of U be unrestricted, but with finite variance, $\mathbb{V}(U) < \infty$.

The sample mean, $\widehat{\beta} = \sum_{i=1}^n U_i/n$, is an asymptotically linear estimator for β with influence function $\phi(U) = U - \mathbb{E}(U)$, knowing that

$$\sqrt{n}(\widehat{\beta} - \beta_0) = n^{-1/2}\sum_{i=1}^n [U_i - \mathbb{E}(U)].$$

Consider any parametric submodel $f(u;\epsilon)$ such that $f(u;\epsilon_0)$ is the true distribution $f_0(u)$. The estimand of interest corresponding to the parametric submodel becomes $\beta(\epsilon) = \int uf(u;\epsilon)du$. Thus,

$$\begin{aligned}\frac{\partial\beta(\epsilon_0)}{\partial\epsilon} &= \int u\frac{\partial f(u;\epsilon_0)}{\partial\epsilon}du = \int u\frac{\partial \log f(u;\epsilon_0)}{\partial\epsilon}f(u;\epsilon_0)du \\ &= \mathbb{E}[US_\epsilon] = \mathbb{E}[(U - \mathbb{E}U)S_\epsilon] = \mathbb{E}[\phi(U)S_\epsilon],\end{aligned}$$

where $S_\epsilon = \partial \log f(u;\epsilon_0)/\partial\epsilon$ is the score. Therefore, by the fundamental theorem of regularity, $\widehat{\beta}$ is regular and therefore it is an RAL estimator.

This simple example has a lot of applications. For example, if the propensity score function, $g_0(A|X)$, is known, then

$$\widehat{\theta}_{\text{rand}} = \frac{1}{n}\sum_{i=1}^n \frac{(2A_i - 1)Y_i}{g_0(A_i|X_i)}$$

is an RAL estimator for the following estimand,

$$\theta_0 = \mathbb{E}\left[\frac{(2A-1)Y}{g_0(A|X)}\right].$$

Here subscript "rand" stands for randomization, because the propensity score function is known in randomized studies. Then it implies that the influence function is

$$\phi_{\text{rand}}(O) = \frac{(2A-1)Y}{g_0(A|X)} - \theta_0.$$

8.2.3 Influence function of MLE-SL estimator

Assume that $\widehat{Q}_{\text{SL}}$ is a consistent estimator of $Q_0(x,a)$. Then under any regular parametric submodel $\{f_X(x;\epsilon_1\}, g(a|x;\epsilon_2), f_Y(y|x,a;\epsilon_3)\}$, the limit of $\widehat{Q}_{\text{SL}}$ is $Q(x,a;\epsilon_3)$. Then, under the parametric submodel, the limit of $\widehat{\theta}_{\text{MLE-SL}}$ is

$$\theta(\boldsymbol{\epsilon}) = \mathbb{E}_{\boldsymbol{\epsilon}}\left[Q(X,1;\epsilon_3) - Q(X,0;\epsilon_3)\right]. \tag{8.13}$$

By differentiation under the integral and the chain rule,

$$\begin{aligned}\partial\theta(\boldsymbol{\epsilon})/\partial\boldsymbol{\epsilon}\Big|_{\boldsymbol{\epsilon}=\mathbf{0}} =& \partial\mathbb{E}_{\boldsymbol{\epsilon}}\left[Q_0(X,1) - Q_0(X,0)\right]/\partial\boldsymbol{\epsilon}\Big|_{\boldsymbol{\epsilon}=\mathbf{0}} &(8.14)\\ &+ \partial\mathbb{E}\left[Q(X,1;\epsilon_3) - Q(X,0;\epsilon_3)\right]/\partial\boldsymbol{\epsilon}\Big|_{\boldsymbol{\epsilon}=\mathbf{0}}. &(8.15)\end{aligned}$$

Using the first direct application of the fundamental theorem of regularity, the first term, (8.14), becomes

$$\partial \mathbb{E}_{\boldsymbol{\epsilon}} \left[Q_0(X,1) - Q_0(X,0) \right] / \partial \boldsymbol{\epsilon} \big|_{\boldsymbol{\epsilon}=\mathbf{0}} = \mathbb{E} \left\{ \left[Q_0(X,1) - Q_0(X,0) \right] S_{\boldsymbol{\epsilon}} \right\}$$

where $S_{\boldsymbol{\epsilon}} = (S_{\epsilon_1}, S_{\epsilon_2}, S_{\epsilon_3})^{\mathrm{T}}$. Note that, by the law of iterated expectations,

$$\mathbb{E} \left\{ \left[Q_0(X,1) - Q_0(X,0) \right] S_{\epsilon_2} \right\} = 0,$$
$$\mathbb{E} \left\{ \left[Q_0(X,1) - Q_0(X,0) \right] S_{\epsilon_3} \right\} = 0.$$

For the second term, (8.15), its derivative over ϵ_1 or ϵ_2 is zero and

$$\begin{aligned} &\partial \mathbb{E} \left[Q(X,1;\epsilon_3) - Q(X,0;\epsilon_3) \right] / \partial \epsilon_3 \big|_{\epsilon_3=0} \\ &= \int \left[\int y \frac{\partial f_Y(y|x,1;0)}{\partial \epsilon_3} dy - \int y \frac{\partial f_Y(y|x,0;0)}{\partial \epsilon_3} dy \right] f_{X0}(x) dx. \end{aligned}$$

We can show that

$$\begin{aligned} &\int \int y \frac{\partial f_Y(y|x,1;0)}{\partial \epsilon_3} dy f_{X0}(x) dx \\ &\overset{(a)}{=} \int \int [y - Q_0(x,1)] \frac{\partial f_Y(y|x,1;0)}{\partial \epsilon_3} dy f_{X0}(x) dx \\ &\overset{(b)}{=} \int \int \left\{ \frac{[y - Q_0(x,1)]}{g_0(1|x)} \frac{\partial f_Y(y|x,1;0)}{\partial \epsilon_3} g_0(1|x) + \frac{0}{g_0(0|x)} g_0(0|x) \right\} dy f_{X0}(x) dx \\ &\overset{(c)}{=} \int \int \mathbb{E} \left\{ \frac{A[y - Q_0(x,A)]}{g_0(A|x)} \frac{\partial f_Y(y|x,A;0)}{\partial \epsilon_3} \Big| X = x \right\} dy f_{X0}(x) dx \\ &\overset{(d)}{=} \int \int \mathbb{E} \left\{ \frac{A[y - Q_0(x,A)]}{g_0(A|x)} \frac{\partial \log f_Y(y|x,A;0)}{\partial \epsilon_3} \Big| X = x \right\} f_{Y0}(y|x,A) dy f_{X0}(x) dx \\ &\overset{(e)}{=} \int \mathbb{E} \left\{ \int \frac{A[y - Q_0(x,A)]}{g_0(A|x)} \frac{\partial \log f_Y(y|x,A;0)}{\partial \epsilon_3} f_{Y0}(y|x,A) dy \Big| X = x \right\} f_{X0}(x) dx \\ &\overset{(f)}{=} \int \mathbb{E} \left\{ \mathbb{E} \left[\frac{A[Y - Q_0(x,A)]}{g_0(A|x)} S_{\epsilon_3} \Big| A, X = x \right] \Big| X = x \right\} f_{X0}(x) dx \\ &\overset{(g)}{=} \int \mathbb{E} \left\{ \frac{A[Y - Q_0(x,A)]}{g_0(A|x)} S_{\epsilon_3} \Big| X = x \right\} f_{X0}(x) dx \\ &\overset{(h)}{=} \mathbb{E} \left\{ \frac{A[Y - Q_0(X,A)]}{g_0(A|X)} S_{\epsilon_3} \right\}, \end{aligned}$$

where (a)–(b) hold because the newly added terms equal zero, (c) holds by writing summation as expectation, (d) holds by forming a score function, (e) holds after switching expectation and integration, (f) holds by writing integration as expectation, (g) holds by the law of iterated expectations, and (h) holds by writing integration as expectation. Similarly, we can show that

$$\int \int y \frac{\partial f_Y(y|x,0;0)}{\partial \epsilon_3} dy f_{X0}(x) dx = \mathbb{E} \left\{ \frac{(1-A)[Y - Q_0(X,A)]}{g_0(A|X)} S_{\epsilon_3} \right\}.$$

Thus,

$$\partial \mathbb{E}\left[Q(X,1;\epsilon_3) - Q(X,0;\epsilon_3)\right]/\partial\epsilon_3\Big|_{\epsilon_3=0} = \mathbb{E}\left\{\frac{(2A-1)[Y-Q_0(X,A)]}{g_0(A|X)}S_{\epsilon_3}\right\}.$$

In addition, because $\mathbb{E}\{[Q(X,1)-Q(X,0)]S_{\epsilon_3}\}=0$ using the law of iterated expectations, we have

$$\begin{aligned}&\partial \mathbb{E}\left[Q(X,1;\epsilon_3) - Q(X,0;\epsilon_3)\right]/\partial\epsilon_3\Big|_{\epsilon_3=0}\\ =&\mathbb{E}\left\{\left[Q(X,1)-Q(X,0)+\frac{(2A-1)[Y-Q_0(X,A)]}{g_0(A|X)}-\theta_0\right]S_{\epsilon_3}\right\}.\end{aligned}$$

And because $\mathbb{E}\{(2A-1)[Y-Q_0(X,A)]/g_0(A|X)S_{\epsilon_1}\}$=0 also using the law of iterated expectations, we have

$$\begin{aligned}&\mathbb{E}\left\{\left[Q_0(X,1)-Q_0(X,0)\right]S_{\epsilon_1}\right\}\\ =&\mathbb{E}\left\{\left[Q(X,1)-Q(X,0)+\frac{(2A-1)[Y-Q_0(X,A)]}{g_0(A|X)}-\theta_0\right]S_{\epsilon_1}\right\}.\end{aligned}$$

Moreover, by the law of iterated expectations, we can verify that

$$\mathbb{E}\left\{\left[Q(X,1)-Q(X,0)+\frac{(2A-1)[Y-Q_0(X,A)]}{g_0(A|X)}-\theta_0\right]S_{\epsilon_2}\right\}=0.$$

Therefore, combining the above three results, we have

$$\frac{\partial\theta(\mathbf{0})}{\partial\epsilon}=\mathbb{E}\left\{\left[Q(X,0)-Q(X,0)+\frac{(2A-1)[Y-Q_0(X,A)]}{g_0(A|X)}-\theta_0\right]S_{\epsilon}\right\}=0,$$

for any regular parametric submodels. Thus, by the fundamental theorem of regularity, we show that the influence function of $\widehat{\theta}_{\text{MLE-SL}}$ is

$$\phi_{\text{MLE-SL}}(O)=Q(X,0)-Q(X,0)+\frac{(2A-1)[Y-Q_0(X,A)]}{g_0(A|X)}-\theta_0. \qquad (8.16)$$

8.2.4 Influence function of IPW-SL estimator

Assume that $\widehat{g}_{\text{SL}}$ is a consistent estimator of $g_0(a|x)$. Then under any regular parametric submodel $\{f_X(x;\epsilon_1\},g(a|x;\epsilon_2),f_Y(y|x,a;\epsilon_3)\}$, the limit of $\widehat{g}_{\text{SL}}$ is $g(a|x;\epsilon_2)$. Then, under the parametric submodel, the limit of $\widehat{\theta}_{\text{IPW-SL}}$ is

$$\theta(\epsilon)=\mathbb{E}_{\epsilon}\left[\frac{(2A-1)Y}{g(A|X;\epsilon_2)}\right]. \qquad (8.17)$$

By differentiation under the integral and the chain rule,

$$\frac{\partial\theta(\epsilon)}{\partial\epsilon}\Big|_{\epsilon=\mathbf{0}}=\partial\mathbb{E}_{\epsilon}\left[\frac{(2A-1)Y}{g_0(A|X)}\right]/\partial\epsilon\Big|_{\epsilon=\mathbf{0}}+\partial\mathbb{E}\left[\frac{(2A-1)Y}{g(A|X;\epsilon_2)}\right]/\partial\epsilon\Big|_{\epsilon=\mathbf{0}}. \qquad (8.18)$$

Using the direct application of the fundamental theorem of regularity, the first term on the right-hand-side (RHS) of (8.18) becomes

$$\partial \mathbb{E}_{\epsilon}\left[\frac{(2A-1)Y}{g_0(A|X)}\right]/\partial \epsilon\Big|_{\epsilon=0} = \mathbb{E}\left[\frac{(2A-1)Y}{g_0(A|X)}S_{\epsilon}\right].$$

For the second term on RHS of (8.18), the derivatives over ϵ_1 are ϵ_3 are equal to zero, respectively, and

$$\begin{aligned}
\partial \mathbb{E}\left[\frac{(2A-1)Y}{g(A|X;\epsilon_2)}\right]/\partial \epsilon_2\Big|_{\epsilon_2=0} &= -\mathbb{E}\left[\frac{(2A-1)Y}{g_0^2(A|X)}\frac{\partial g(A|X;0)}{\partial \epsilon_2}\right] \\
&= -\mathbb{E}\left[\frac{(2A-1)Y}{g_0(A|X)}\frac{\partial \log g(A|X;0)}{\partial \epsilon_2}\right] \\
&= -\mathbb{E}\left[\frac{(2A-1)Y}{g_0(A|X)}S_{\epsilon_2}\right].
\end{aligned}$$

Combining these two results, we have

$$\frac{\partial \theta(\mathbf{0})}{\partial \epsilon} = \left(\mathbb{E}\left[\frac{(2A-1)Y}{g_0(A|X)}S_{\epsilon_1}\right], 0, \mathbb{E}\left[\frac{(2A-1)Y}{g_0(A|X)}S_{\epsilon_3}\right]\right)^{\mathrm{T}}.$$

Therefore, by the first direct application of the fundamental theorem of regularity again, we see that the influence function of $\widehat{\theta}_{\text{IPW-SL}}$ is

$$\phi_{\text{IPW-SL}}(O) = \frac{(2A-1)Y}{g_0(A|X)} + h(X,A) - \theta_0, \tag{8.19}$$

where $h(X,A)$ is a function to be determined, which satisfies

$$\mathbb{E}\left[h(X,A)S_{\epsilon_1}\right] = 0, \tag{8.20}$$

$$\mathbb{E}\left\{\left[\frac{(2A-1)Y}{g_0(A|X)} + h(X,A)\right]S_{\epsilon_2}\right\} = 0, \tag{8.21}$$

$$\mathbb{E}\left[h(X,A)S_{\epsilon_3}\right] = 0, \tag{8.22}$$

for any regular parametric submodels.

First, (8.22) holds for any regular parametric submodels, using the law of iterated expectations and the fact that $\mathbb{E}(S_{\epsilon_3}|X,A) = 0$.

Second, in order to satisfy (8.20) for any regular parametric submodels, it requires that $\mathbb{E}[h(X,A)|X] = 0$. Thus, it requires $h(X,1)g_0(1|X) + h(X,0)[1 - g_0(1|X)] = 0$. That is, it requires $h(X,1) = -h(X,0)/g_0(1|X)[1 - g(1|X)]$, along with an apparent fact that $h(X,0) = -h(X,0)/g_0(1|X)[0 - g(1|X)]$. Therefore, we see that $h(X,A) = \{A - g_0(1|X)\}h^*(X)$, where $h^*(X) = h(X,0)/g_0(1|X)$ is to be determined.

Third, in order to satisfy (8.21),

$$\mathbb{E}\left\{\left[\frac{(2A-1)Y}{g_0(A|X)} + \{A - g_0(1|X)\}h^*(X)\right]S_{\epsilon_2}\right\} = 0,$$

for any regular parametric submodels, it requires that

$$\mathbb{E}\left\{\left[\frac{(2A-1)Y}{g_0(A|X)}+\{A-g_0(1|X)\}h^*(X)\right][\{A-g_0(1|X)\}\alpha_2(X)]\right\}=0,$$

for any $\alpha_2(X)$, using the membership of $\mathcal{T}_2$. Thus, by the law of iterated expectations, it requires that

$$\mathbb{E}\left\{\left[\frac{(2A-1)\{A-g_0(1|X)\}Q_0(X,A)}{g_0(A|X)}+\{A-g_0(1|X)\}^2h^*(X)\right]\alpha_2(X)\right\}=0,$$

for any $\alpha_2(X)$. By the law of iterated expectations again, it requires that

$$\mathbb{E}\left\{[g_0(0|X)Q_0(X,1)+g_0(1|X)Q_0(X,0)+g_0(1|X)g_0(0|X)h^*(X)]\,\alpha_2(X)\right\}=0,$$

for any $\alpha_2(X)$. Therefore, it requires that

$$g_0(0|X)Q_0(X,1)+g_0(1|X)Q_0(X,0)+g_0(1|X)g_0(0|X)h^*(X)=0,$$

leading to

$$h^*(X)=-\left[\frac{Q_0(X,1)}{g_0(1|X)}+\frac{Q_0(X,0)}{g_0(0|X)}\right].$$

Therefore, plugging the resulting $h^*(X)$ into (8.19), we show that

$$\phi_{\text{IPW-SL}}(O)=\frac{(2A-1)Y}{g_0(A|X)}-\{A-g_0(1|X)\}\sum_{j=0}^{1}\frac{Q_0(X,j)}{g_0(j|X)}-\theta_0. \qquad (8.23)$$

Aha! This influence function is the same as the one we developed in the construction of the AIPW estimator in Chapter 7 using only the tools of elementary calculus, without using the fundamental theorem of regularity.

8.2.5 Double robustness of AIPW-SL estimator

The performance of the MLE-SL estimator depends on whether $\widehat{Q}_{\text{SL}}$ is a consistent estimator of Q_0. Thus, the MLE-SL estimator is singly robust. If $\widehat{Q}_{\text{SL}}$ is a consistent estimator of Q_0, we show that the asymptotic variance can be estimated based on the influence function (8.16). But the influence function depends on g_0, which should be estimated, say by $\widehat{g}_{\text{SL}}$.

The performance of the IPW-SL estimator depends on whether $\widehat{g}_{\text{SL}}$ is a consistent estimator of g_0. Thus, the IPW-SL estimator is singly robust. If $\widehat{g}_{\text{SL}}$ is a consistent estimator of g_0, we show that the asymptotic variance can be estimated based on the influence function (8.23). But the influence function depends on Q_0, which should be estimated, say by $\widehat{Q}_{\text{SL}}$.

Since, in order to estimate the asymptotic variance of either MLE-SL or IPW-SL, we need to estimate both g_0 and Q_0 no matter what, then why not consider the AIPW-SL estimator to begin with? The AIPW-SL estimator, (8.10), is doubly robust, in the sense that it is consistent if either $\widehat{g}_{\text{SL}}$ is a consistent estimator of g_0 or $\widehat{Q}_{\text{SL}}$ is a consistent estimator of Q_0.

Simply put, the construction of AIPW-SL depends on $\widehat{g}_{\text{SL}}$ and $\widehat{Q}_{\text{SL}}$, and the asymptotic variance of AIPW-SL can be estimated by

$$\frac{1}{n(n-1)}\sum_{i=1}^{n}\left[\frac{(2A_i-1)Y_i}{\widehat{g}_{\text{SL}}(A_i|X_i)}+\{A_i-\widehat{g}_{\text{SL}}(1|X_i)\}\sum_{j=0}^{1}\frac{\widehat{Q}_{\text{SL}}(X_i,j)}{\widehat{g}_{\text{SL}}(j|X_i)}-\widehat{\theta}_{\text{AIPW-SL}}\right]^2.$$

8.2.6 Efficient influence function

There is a geometrical viewpoint of influence functions (Bickel et al. 1993). Let $\mathcal{H}$ be the Hilbert space of random functions, $h(O)$, with mean zero and finite variance, equipped with the inner product $\langle h_1, h_2\rangle = \mathbb{E}(h_1 h_2)$. Therefore, influence functions are members of $\mathcal{H}$.

Our goal is to identify the "optimal" influence function, referred to as the *efficient influence function*, $\phi_{\text{EIF}}(O) \in \mathcal{H}$, such that the variance of an RAL estimator with $\phi_{\text{EIF}}(O)$ achieves the Cramer-Rao bound. The efficient influence function can be expressed as

$$\phi_{\text{EIF}} = \arg\min_{\phi\in\mathcal{S}_{\text{RAL}}} \mathbb{V}(\phi(O)), \tag{8.24}$$

where $\mathcal{S}_{\text{RAL}}$ is the set of all the influence functions associated with RAL estimators for θ.

The equivalence between the above two influence functions

We have shown that the influence functions of the MLE-SL estimator and the IPW-SL estimator are, respectively,

$$Q_0(X,0) - Q_0(X,0) + \frac{(2A-1)}{g_0(A|X)}[Y - Q_0(X,A)] - \theta_0. \tag{8.25}$$

and

$$\frac{(2A-1)Y}{g_0(A|X)} - \left[\frac{Q_0(X,1)}{g_0(1|X)} + \frac{Q_0(X,0)}{g_0(0|X)}\right][A - g_0(1|X)] - \theta_0. \tag{8.26}$$

These two influence functions are equal—they are just two forms of the same thing. To show the above two forms, (8.25) and (8.26), are equivalent, we check the case where $A = 1$ and the case where $A = 0$, respectively. In fact, when $A = 1$, both (8.25) and (8.26) become

$$\frac{Y}{g_0(1|X)} - Q_0(X,1)\frac{g_0(0|X)}{g_0(1|X)} - Q_0(X,0) - \theta_0,$$

while when $A = 0$, both (8.25) and (8.26) become

$$-\frac{Y}{g_0(0|X)} + Q_0(X,1) + Q_0(X,0)\frac{g_0(1|X)}{g_0(0|X)} - \theta_0.$$

The fact that the above two influence functions are the same is not surprising because we will soon show that, when there is no constraint on f_{X0}, g_0, and f_{Y0}, all the RAL estimators have the same influence function; that is, there is only one influence function that satisfies the fundamental theorem of regularity, if there is no constraint on the underlying distributions. Therefore, we will call this influence function the *efficient influence function*:

$$\begin{aligned}\phi_{\text{EIF}}(O) &= Q_0(X,1) - Q_0(X,0) + \frac{(2A-1)}{g_0(A|X)}[Y - Q_0(X,A)] - \theta_0 \\ &= \frac{(2A-1)Y}{g_0(A|X)} - \left[\frac{Q_0(X,1)}{g_0(1|X)} + \frac{Q_0(X,0)}{g_0(0|X)}\right][A - g_0(1|X)] - \theta_0. \quad (8.27)\end{aligned}$$

The efficient influence function

For any function $h(O) \in \mathcal{H}$, we have the following decomposition,

$$\begin{aligned}h(O) &= \mathbb{E}[h(O)|X] + \{\mathbb{E}[h(O)|X,A] - \mathbb{E}[h(O)|X]\} + \{h(O) - \mathbb{E}[h(O)|X,A]\} \\ &= \alpha_1(X) + \alpha_2(X,A) + \alpha_3(X,A,Y),\end{aligned}$$

where $\mathbb{E}[\alpha_1(X)] = 0$, $\mathbb{E}[\alpha_2(X,A)|X] = 0$, and $\mathbb{E}[\alpha_3(X,A,Y)|X,A] = 0$. Hence,

$$\mathcal{H} = \mathcal{T}_1 \oplus \mathcal{T}_2 \oplus \mathcal{T}_3. \quad (8.28)$$

Previously we have shown that $\phi_{\text{EIF}}(O)$ satisfies the fundamental theorem of regularity; that is,

$$\partial\theta(\epsilon)/\partial\epsilon\big|_{\epsilon=0} = \mathbb{E}[\phi_{\text{EIF}}(O)S_\epsilon].$$

Assume there is another influence function $\widetilde{\phi}(O)$ that also satisfies the fundamental theorem of regularity,

$$\partial\theta(\epsilon)/\partial\epsilon\big|_{\epsilon=0} = \mathbb{E}[\widetilde{\phi}(O)S_\epsilon].$$

Thus, $\mathbb{E}\{[\widetilde{\phi}(O) - \phi_{\text{EIF}}(O)]S_\epsilon]\} = 0$ for all regular parametric submodels. By (8.28), $\mathbb{E}\{[\widetilde{\phi}(O) - \phi_{\text{EIF}}(O)]h(O)\} = 0$, for any $h \in \mathcal{H}$, implying $\widetilde{\phi}(O) = \phi_{\text{EIF}}(O)$. Therefore, there is only one influence function that satisfies the fundamental theorem of regularity and this influence function achieves the minimum of (8.24) automatically.

8.3 The Targeted Learning Framework

In this section, we describe the targeted learning framework briefly. Refer to van der Laan and Rose (2011) for more detail.

8.3.1 Mini-roadmap

A full roadmap for conducting causal inference will be discussed in Chapter 11. In this subsection, we provide a mini-roadmap of using the targeted learning approach to estimate the ATE estimand.

Step 1: The ATE estimand

Assume we target at the following ATE estimand,

$$\theta = \theta(Q, F_X) = \int [Q(x,1) - Q(x,0)]dF_X(x). \tag{8.29}$$

Step 2: Initial estimators of Q and g

Obtain an initial estimator, $\widehat{Q}_{\text{SL}}$, for the outcome regression function Q. Also obtain an initial estimator, $\widehat{g}_{\text{SL}}$, for the propensity score function g.

Step 3: The efficient influence function of the ATE estimand

Derive the following efficient influence function,

$$\phi_{\text{EIF}}(O) = Q_0(X,1) - Q_0(X,0) + \frac{(2A-1)}{g_0(A|X)}[Y - Q_0(X,A)] - \theta_0.$$

Step 4: TMLE

Update the initial estimator to obtain the targeted maximum likelihood estimator/targeted minimum loss estimator (TMLE) using the procedure to be described in the next section.

8.3.2 TMLE

We start with the following initial estimator for θ,

$$\widehat{\theta}_{\text{MLE-SL}} = \theta(\widehat{Q}_{\text{SL}}, \widehat{F}_X) = \frac{1}{n}\sum_{i=1}^{n}\left[\widehat{Q}_{\text{SL}}(X_i,1) - \widehat{Q}_{\text{SL}}(X_i,0)\right].$$

Then, using the efficient influence function, we are able to update the initial estimator $\widehat{Q}_{\text{SL}}$ to $\widehat{Q}^*$, producing an updated estimator for the target estimand (van der Laan and Rose 2011), which is referred to as TMLE, denoted as $\widehat{\theta}_{\text{SL-MLE}}$; that is,

$$\widehat{\theta}_{\text{TMLE}} = \theta(\widehat{Q}^*, \widehat{F}_X) = \frac{1}{n}\sum_{i=1}^{n}\left[\widehat{Q}^*(X_i,1) - \widehat{Q}^*(X_i,0)\right]. \tag{8.30}$$

van der Laan and Rose (2011) showed that $\widehat{\theta}_{\mathrm{TMLE}}$ is an RAL estimator with the efficient influence function; that is,

$$\sqrt{n}(\widehat{\theta}_{\mathrm{TMLE}} - \theta_0) = \frac{1}{\sqrt{n}} \sum_{i=1}^{n} \phi_{\mathrm{eff}}(O_i) + o_p(1). \tag{8.31}$$

At first glance, TMLE seems mysterious. But van der Laan and Rose (2011) and van der Laan and Rubin (2006) proposed a clever way to update the estimator of Q from $\widehat{Q}_{\mathrm{SL}}$ to $\widehat{Q}^*$. Here we discuss their method briefly in three steps; refer to van der Laan and Rose (2011) for more detail.

First, define the so-called "clever covariate",

$$H^*(X, A) = \frac{2A - 1}{\widehat{g}_{\mathrm{SL}}(A|X)}. \tag{8.32}$$

Note the clever covariate appears in the efficient influence function (8.30) as the coefficient of $Y - Q_0(X, A)$.

Second, run a logistic regression—referred to as "clever logistic regression", in which the outcome variable is either binary Y or bounded continuous variable (if $Y \in [a, b]$, then $\widetilde{Y} = (Y - a)/(b - a) \in [0, 1]$), the intercept is set as $\mathrm{logit}[\widehat{Q}_{\mathrm{SL}}(X, A)]$, and there is only one covariate which is the clever covariate $H^*(X, A)$. Obtain the resulting coefficient in front of the clever covariate, which is the maximizer of the likelihood of the clever logistic regression,

$$\widehat{\epsilon}_3 = \arg\max_{\epsilon_3} \prod_{i=1}^{n} \frac{\left[\exp\{\mathrm{logit}[\widehat{Q}_{\mathrm{SL}}(X_i, A_i)] + \epsilon_3 H^*(X_i, A_i)\}\right]^{Y_i}}{1 + \exp\{\mathrm{logit}[\widehat{Q}_{\mathrm{SL}}(X_i, A_i)] + \epsilon_3 H^*(X_i, A_i)\}}. \tag{8.33}$$

Third, obtain the following updated estimator of Q,

$$\widehat{Q}^*(X, A) = \frac{\exp\{\mathrm{logit}[\widehat{Q}_{\mathrm{SL}}(X, A)] + \widehat{\epsilon}_3 H^*(X, A)\}}{1 + \exp\{\mathrm{logit}[\widehat{Q}_{\mathrm{SL}}(X, A)] + \widehat{\epsilon}_3 H^*(X, A)\}}. \tag{8.34}$$

8.3.3 Double robustness

Like AIPW-SL, TMLE is doubly robust in the sense that $\widehat{\theta}_{\mathrm{TMLE}}$ is a consistent estimator of θ_0 if either $\widehat{Q}_{\mathrm{SL}}$ is a consistent estimator of Q_0 or $\widehat{g}_{\mathrm{SL}}$ is a consistent estimator of g_0.

Moreover, in finite sample, TMLE performs better than AIPW-SL in the sense that TMLE is more stable because it is a plug-in estimator whereas AIPW-SL is not. Being a plug-in estimator, the value of the estimator stays in the range of the estimand. Refer to van der Laan and Rose (2011) for more comparisons between the two methods.

The asymptotic variance of TMLE can be estimated by

$$\frac{1}{n(n-1)} \sum_{i=1}^{n} \left[\widehat{Q}^*(X_i, 1) - \widehat{Q}^*(X_i, 0) + \frac{(2A_i - 1)}{\widehat{g}_{\mathrm{SL}}(A_i|X_i)}[Y_i - \widehat{Q}^*(X_i, A_i)] - \widehat{\theta}_{\mathrm{TMLE}}\right]^2.$$

8.4 A Shortcut to Derive Efficient Influence Functions

The construction of TMLE depends on knowing the efficient influence function. We have seen how to use the fundamental theorem of regularity to find the efficient influence function for estimating the ATE estimand. But the process of finding the efficient influence function is tedious. In this section, we provide a shortcut to derive the efficient influence function and apply this shortcut to find the efficient influence function for estimating the ATT estimand and for estimating the ATE estimand when there are missing data.

8.4.1 The efficient influence function for ATE

We want to estimate the average treatment effect (ATE),

$$\theta^*_{\text{ATE}} = \mathbb{E}(Y^{a=1} - Y^{a=0}),$$

and we have data $\{O_i = (X_i, A_i, Y_i), i = 1, \ldots, n\}$.

Under the identifiability assumptions (consistency, exchangeability, and positivity), there are two strategies—the standardization strategy and the weighting strategy—to translate the causal estimand into the following statistical estimand with two different forms:

$$\theta_{\text{ATE}} = \mathbb{E}[Q_0(X,1) - Q_0(X,0)] = \mathbb{E}\left[\frac{(2A-1)Y}{g_0(A|X)}\right].$$

Here is the shortcut to find the efficient influence function:

- Step 1: Let

$$\phi_1(O) = Q_0(X,1) - Q_0(X,0) - \theta_{\text{ATE}},$$

which is the centralized term inside the expectation in the statistical estimand defined via the standardization strategy;

- Step 2: Let

$$\phi_2(O) = \frac{(2A-1)[Y - Q_0(A|X)]}{g_0(A|X)},$$

which is the centralized term inside the expectation in the statistical estimand defined via the weighting strategy;

- Step 3: The efficient influence function is $\phi_{\text{EIF-ATE}}(O) = \phi_1(0) + \phi_2(O)$,

$$\phi_{\text{EIF-ATE}}(O) = \frac{(2A-1)[Y - Q_0(A|X)]}{g_0(A|X)} + Q_0(X,1) - Q_0(X,0) - \theta_{\text{ATE}}.$$

Proof of the validity of the shortcut:

For any given regular parametric submodel indicated by $\boldsymbol{\epsilon} = (\epsilon_1, \epsilon_2, \epsilon_3)^{\mathrm{T}}$, where ϵ_1 is for $f_X(x)$, ϵ_2 is for $g(a|x)$, and ϵ_3 is for $f_Y(y|x,a)$, the statistical estimand becomes a functional of $\boldsymbol{\epsilon}$ with two different forms:

$$\theta_{\mathrm{ATE}}(\boldsymbol{\epsilon}) = \mathbb{E}_{\boldsymbol{\epsilon}}[Q(X,1;\epsilon_3) - Q(X,0;\epsilon_3)] = \mathbb{E}_{\boldsymbol{\epsilon}}\left[\frac{(2A-1)Y}{g(A|X;\epsilon_2)}\right].$$

We can verify that the function $\phi_{\text{EIF-ATE}}(O)$ defined via the shortcut satisfies the fundamental theorem of regularity condition,

$$\partial\theta_{\mathrm{ATE}}(\boldsymbol{\epsilon})/\partial\boldsymbol{\epsilon}\big|_{\boldsymbol{\epsilon}=\mathbf{0}} = \mathbb{E}[\phi_{\text{EIF-ATE}}(O)S_{\boldsymbol{\epsilon}}],$$

which is the same as

$$\partial\theta_{\mathrm{ATE}}(\boldsymbol{\epsilon})/\partial\epsilon_j|_{\boldsymbol{\epsilon}=\mathbf{0}} = \mathbb{E}\left\{[\phi_1(O)+\phi_2(O)]S_{\epsilon_j}]\right\}, j=1,2,3.$$

In fact, for $j=1$ or 2, because

$$\partial\mathbb{E}_{\boldsymbol{\epsilon}}[Q(X,1;\epsilon_3) - Q(X,0;\epsilon_3)]/\partial\epsilon_j|_{\boldsymbol{\epsilon}=\mathbf{0}} = \mathbb{E}[\phi_1(O)S_{\epsilon_j}]$$

using the fact that $Q(X,1;\epsilon_3) - Q(X,0;\epsilon_3)$ does not depend on ϵ_j and because

$$\mathbb{E}[\phi_2(O)S_{\epsilon_j}] = \mathbb{E}\{\mathbb{E}[\phi_2(O)S_{\epsilon_j}|A,X]\} = \mathbb{E}\{S_{\epsilon_j}\mathbb{E}[\phi_2(O)|A,X]\} = 0$$

using the law of iterated expectations, we have

$$\partial\theta_{\mathrm{ATE}}(\boldsymbol{\epsilon})/\partial\epsilon_j|_{\boldsymbol{\epsilon}=\mathbf{0}} = \mathbb{E}\left\{[\phi_1(O)+\phi_2(O)]S_{\epsilon_j}\right\}, j=1,2. \tag{8.35}$$

For $j=3$, because

$$\partial\mathbb{E}_{\boldsymbol{\epsilon}}\left[\frac{(2A-1)Y}{g(A|X;\epsilon_2)}\right]\Big/\partial\epsilon_3\Big|_{\boldsymbol{\epsilon}=\mathbf{0}} = \mathbb{E}[\phi_2(O)S_{\epsilon_3}]$$

using the fact that $(2A-1)Y/g(A|X;\epsilon_2)$ does not depends on ϵ_3 and because

$$\mathbb{E}[\phi_1(O)S_{\epsilon_3}] = \mathbb{E}\{\mathbb{E}[\phi_1(O)S_{\epsilon_3}|A,X]\} = \mathbb{E}\{\phi_1(O)\mathbb{E}[S_{\epsilon_3}|A,X]\} = 0$$

using the law of iterated expectations, we have

$$\partial\theta_{\mathrm{ATE}}(\boldsymbol{\epsilon})/\partial\epsilon_3|_{\boldsymbol{\epsilon}=\mathbf{0}} = \mathbb{E}\left\{[\phi_1(O)+\phi_2(O)]S_{\epsilon_2}\right\}. \tag{8.36}$$

Thus, combining results (8.35) and (8.36), we complete the proof. □

8.4.2 The efficient influence function for ATT

We want to estimate the average treatment effect on the treated (ATT),

$$\theta^*_{\text{ATT}} = \mathbb{E}(Y^{a=1} - Y^{a=0}|A = 1),$$

and we have data $\{O_i = (X_i, A_i, Y_i, i = 1, \ldots, n\}$.

Under the identifiability assumptions (consistency, exchangeability, and positivity), there are two strategies to translate the causal estimand into the following statistical estimand with two different forms:

$$\theta_{\text{ATT}} = \mathbb{E}\left\{\frac{I(A=1)}{\mathbb{P}(A=1)}[Q_0(X,1) - Q_0(X,0)]\right\} \tag{8.37}$$

by the standardization strategy, and

$$\theta_{\text{ATT}} = \mathbb{E}\left\{\left[\frac{I(A=1)}{\mathbb{P}(A=1)} - \frac{I(A=0)g_0(1|X)}{\mathbb{P}(A=1)g_0(0|X)}\right]Y\right\}, \tag{8.38}$$

by the weighting strategy.

Thus, we can apply the shortcut to find the efficient influence function:

- Step 1: Let

$$\phi_1(O) = \frac{I(A=1)}{\mathbb{P}(A=1)}[Q_0(X,1) - Q_0(X,0)] - \theta_{\text{ATT}},$$

  which is the centralized term inside the expectation in (8.37);

- Step 2: Let

$$\phi_2(O) = \left[\frac{I(A=1)}{\mathbb{P}(A=1)} - \frac{I(A=0)g_0(1|X)}{\mathbb{P}(A=1)g_0(0|X)}\right][Y - Q_0(X,A)],$$

  which is the centralized term inside the expectation in (8.38);

- Step 3: The efficient influence function is $\phi_{\text{EIF-ATT}}(O) = \phi_1(0) + \phi_2(O)$,

$$\begin{aligned}\phi_{\text{EIF-ATT}}(O) = &\left[\frac{I(A=1)}{\mathbb{P}(A=1)} - \frac{I(A=0)g_0(1|X)}{\mathbb{P}(A=1)g_0(0|X)}\right][Y - Q_0(X,A)] \\ &+ \frac{I(A=1)}{\mathbb{P}(A=1)}[Q_0(X,1) - Q_0(X,0)] - \theta_{\text{ATT}}.\end{aligned}$$

We can follow the similar statements as those in the previous subsection to prove the validity of the shortcut.

8.4.3 Missing data due to analysis dropout

Assume we apply the hypothetical strategy—envisage a hypothetical scenario in which no analysis dropout would occur—to handle missing data due to analysis dropouts. That is, we want to estimate the following estimand,

$$\theta^*_{\mathrm{H}} = \mathbb{E}(Y^{a=1,\delta=0} - Y^{a=0,\delta=0}),$$

and we have data $\{O_i = (X_i, A_i, \Delta_i, (1-\Delta_i)Y_i), i = 1, \ldots, n\}$.

In Chapter 3, we showed that under the identifiability assumptions (consistency, exchangeability, missing at random, and positivity), by the standardization strategy (although we were not aware of the name then), we can translate the causal estimand into the following statistical estimand,

$$\theta_{\mathrm{H}} = \mathbb{E}[\widetilde{Q}_0(X,1) - \widetilde{Q}_0(X,0)], \tag{8.39}$$

where $\widetilde{Q}_0(X,A) = \mathbb{E}(Y|X,A,\Delta=0)$. Note that here $\widetilde{Q}$ is the same as $Q_{\Delta=0}$ defined in Chapter 3.

Alternatively, if we think of "$A=1, \Delta=0$" and "$A=0, \Delta=0$" as actions parallel to "$A=1$" and "$A=0$", by the weighting strategy, we can show that the above statistical estimanid can also be expressed as,

$$\begin{aligned}
\theta_{\mathrm{H}} &= \mathbb{E}\left[\frac{I(A=1,\Delta=0)Y}{\mathbb{P}(A=1,\Delta=0|X)} - \frac{I(A=0,\Delta=0)Y}{\mathbb{P}(A=0,\Delta=0|X)}\right] \\
&= \mathbb{E}\left[\frac{I(A=1,\Delta=0)Y}{\mathbb{P}(A=1|X)\mathbb{P}(\Delta=0|X,A=1)} - \frac{I(A=0,\Delta=0)Y}{\mathbb{P}(A=0|X)\mathbb{P}(\Delta=0|X,A=0)}\right] \\
&= \mathbb{E}\left[\frac{I(A=1,\Delta=0)Y}{g_0(1|X)h_0(X,1)} - \frac{I(A=0,\Delta=0)Y}{g_0(0|X)h_0(X,0)}\right], \\
&= \mathbb{E}\left\{\left[\frac{I(A=1,\Delta=0)}{g_0(1|X)h_0(X,1)} - \frac{I(A=0,\Delta=0)}{g_0(0|X)h_0(X,0)}\right]Y\right\},
\end{aligned} \tag{8.40}$$

where $g_0(a|x) = \mathbb{P}(A=a|X=x)$ is the propensity score function and $h_0(x,a) = \mathbb{P}(\Delta=0|X=x,A=a)$ is the non-missing probability function.

Thus, we obtain two equivalent forms of the same statistical estimand θ_H. Here is the shortcut to find the efficient influence function:

- Step 1: Let

  $$\phi_1(O) = \widetilde{Q}_0(X,1) - \widetilde{Q}_0(X,0) - \theta_{\mathrm{H}},$$

  which is the centralized term inside the expectation in (8.39);

- Step 2: Let

  $$\phi_2(O) = \left[\frac{I(A=1,\Delta=0)}{g_0(1|X)h_0(X,1)} - \frac{I(A=0,\Delta=0)}{g_0(0|X)h_0(X,0)}\right][Y - \widetilde{Q}_0(X,A)],$$

  which is the centralized term inside the expectation in (8.40);

- Step 3: The efficient influence function is $\phi_{\text{EIF-H}}(O) = \phi_1(0) + \phi_2(O)$,

$$\begin{aligned}\phi_{\text{EIF-H}}(O) = &\left[\frac{I(A=1,\Delta=0)}{g_0(1|X)h_0(X,1)} - \frac{I(A=0,\Delta=0)}{g_0(0|X)h_0(X,0)}\right][Y - \widetilde{Q}_0(X,A)] \\ &+ \widetilde{Q}_0(X,1) - \widetilde{Q}_0(X,0) - \theta_{\text{H}}.\end{aligned}$$

Proof of the validity of the shortcut:

Consider any regular parametric submodel indicated by $\boldsymbol{\epsilon} = (\epsilon_1, \epsilon_2, \epsilon_3, \epsilon_4)^{\text{T}}$, where $f_X(x; \epsilon_1)$ is equal to $f_{X0}(X)$ when $\epsilon_1 = 0$, $g(a|x; \epsilon_2)$ is equal to $g_0(a|x)$ when $\epsilon_2 = 0$, $h(x, a; \epsilon_3)$ is equal to $h_0(x, a)$ when $\epsilon_3 = 0$, and $f_Y(y|x, a, \delta = 0; \epsilon_4)$ is equal to $f_{Y0}(y|x, a, \delta = 0)$ when $\epsilon_4 = 0$. Thus, under the regular parametric submodel, the statistical estimand becomes a function of ϵ with two equivalent forms:

$$\theta_{\text{H}}(\boldsymbol{\epsilon}) = \mathbb{E}_{\boldsymbol{\epsilon}}[\widetilde{Q}(X, 1; \epsilon_4) - \widetilde{Q}(X, 0; \epsilon_4)] \tag{8.41}$$

by the standardization strategy, and

$$\theta_{\text{H}}(\boldsymbol{\epsilon}) = \mathbb{E}_{\boldsymbol{\epsilon}}\left\{\left[\frac{I(A=1,\Delta=0)}{g(1|X;\epsilon_2)h(X,1;\epsilon_3)} - \frac{I(A=0,\Delta=0)}{g(0|X;\epsilon_2)h(X,0;\epsilon_3)}\right]Y\right\} \tag{8.42}$$

by the weighting strategy.

We can verify that function $\phi_{\text{EIF-H}}(O)$ defined via the shortcut satisfies the the fundamental theorem of regularity condition,

$$\partial\theta_{\text{H}}(\boldsymbol{\epsilon})/\partial\boldsymbol{\epsilon}\big|_{\boldsymbol{\epsilon}=\mathbf{0}} = \mathbb{E}[\phi_{\text{EIF-H}}(O)S_{\boldsymbol{\epsilon}}],$$

which is the same as

$$\partial\theta_{\text{H}}(\boldsymbol{\epsilon})/\partial\epsilon_j|_{\boldsymbol{\epsilon}=\mathbf{0}} = \mathbb{E}\left\{[\phi_1(O) + \phi_2(O)]S_{\epsilon_j}]\right\}, j = 1, 2, 3, 4.$$

In fact, for $j = 1, 2, 3$, because

$$\partial\mathbb{E}_{\boldsymbol{\epsilon}}[\widetilde{Q}(X, 1; \epsilon_4) - \widetilde{Q}(X, 0; \epsilon_4)]/\partial\epsilon_j|_{\boldsymbol{\epsilon}=\mathbf{0}} = \mathbb{E}[\phi_1(O)S_{\epsilon_j}]$$

by the fact that $\widetilde{Q}(X, 1; \epsilon_4) - \widetilde{Q}(X, 0; \epsilon_4)$ does not depend on ϵ_j and because

$$\mathbb{E}[\phi_2(O)S_{\epsilon_j}] = \mathbb{E}\left\{S_{\epsilon_j}\mathbb{E}[\phi_2(O)|X, A, \Delta = 0]\right\} = 0$$

using the law of iterated expectations, we have

$$\partial\theta_{\text{H}}(\boldsymbol{\epsilon})/\partial\epsilon_j\big|_{\boldsymbol{\epsilon}=\mathbf{0}} = \mathbb{E}\left\{[\phi_1(O) + \phi_2(O)]S_{\epsilon_j}\right\}, j = 1, 2, 3.$$

For $j = 4$, because

$$\frac{\partial\mathbb{E}_{\boldsymbol{\epsilon}}}{\partial\epsilon_4}\left\{\left[\frac{I(A=1,\Delta=0)}{g(1|X;\epsilon_2)h(X,1;\epsilon_3)} - \frac{I(A=0,\Delta=0)}{g(0|X;\epsilon_2)h(X,0;\epsilon_3)}\right]Y\right\}\Bigg|_{\boldsymbol{\epsilon}=\mathbf{0}} = \mathbb{E}[\phi_2(O)S_{\epsilon_4}]$$

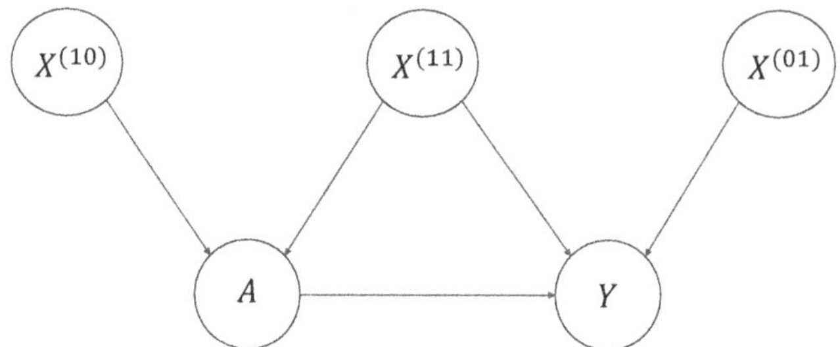

FIGURE 8.1
Selection of covariates

by the fact that the term inside the expectation on the left-hand-side of the above equation does not depend on ϵ_4 and because

$$\mathbb{E}[\phi_1(O)S_{\epsilon_4}] = \mathbb{E}\left\{\phi_1(O)\mathbb{E}[S_{\epsilon_4}|A, X, \Delta = 0]\right\} = 0$$

using the law of iterated expectations, we have

$$\partial\theta_{\text{H}}(\epsilon)/\partial\epsilon_4|_{\boldsymbol{\epsilon}=\mathbf{0}} = \mathbb{E}\left\{[\phi_1(O) + \phi_2(O)]S_{\epsilon_4}\right\}.$$

Thus, we complete the proof. □

8.5 Discussion

8.5.1 How to select covariates?

As shown in Figure 8.1, which was motivated by Figure 1 of Brookhart et al. (2006), $X^{(11)}$ is the vector of covariates that are associated with both A and Y (i.e., the set of confounders that ensure the exchangeability assumption be satisfied), $X^{(10)}$ is the vector of covariates associated with A but not Y, and $X^{(01)}$ is the vector of covariates associated with Y but not A.

Using simulation studies, Brookhart et al. (2006) demonstrated that we'd better consider $X = (X^{(11)}, X^{(01)})$, which consists of all the covariates that are believed to be associated with Y. Their simulation studies demonstrated that including confounders $X^{(11)}$ helps with the adjustment of confounding bias and including $X^{(01)}$ helps with the reduction of the variance of the estimator.

In this subsection, we will use the variance of the efficient influence function $\phi_{\text{EIF-ATE}}(O)$ to show that why we should consider $X = (X^{(11)}, X^{(01)})$. For this aim, we consider the following decomposition,

$$\mathbb{V}[\phi_{\text{EIF-ATE}}(O)] = \mathbb{E}\left\{\mathbb{V}[\phi_{\text{EIF-ATE}}|A, X]\right\} + \mathbb{V}\left\{\mathbb{E}[\phi_{\text{EIF-ATE}}|A, X]\right\}.$$

Note that

$$\mathbb{E}\{\mathbb{V}[\phi_{\text{EIF-ATE}}|A,X]\} = \mathbb{E}\left\{\left[\frac{(2A-1)}{g_0(A|X)}\right]^2 \mathbb{V}[Y - Q_0(A,X)|A,X]\right\}, \quad (8.43)$$

$$\mathbb{V}\{\mathbb{E}[\phi_{\text{EIF-ATE}}|A,X]\} = \mathbb{V}[Q_0(X,1) - Q_0(X,0) - \theta_{\text{ATE}}]. \quad (8.44)$$

First, X should include $X^{(11)}$ to adjust for confounding bias. Second, X may include $X^{(01)}$ to reduce $\mathbb{V}[Y - Q_0(A,X)|A,X]$ in (8.43). Third, X shouldn't include $X^{(10)}$ because including it wouldn't help reduce either $\mathbb{V}[Y - Q_0(A,X)|A,X]$ in (8.43) or $\mathbb{V}[Q_0(X,1) - Q_0(X,0)]$ in (8.44), but including $X^{(10)}$ would make $g_0(A|X)$ closer to 1 or 0, making $[(2A-1)/g_0(A|X)]^2$ in (8.43) more unstable.

8.5.2 How to handle missing covariates?

In practice, there may be missing data in some covariates—to be more accurate, there may be non-available (NA) values in some covariates. To solve the problem of NA values in the covariates, we expand the covariate vector by adding one indicator variable for each covariate that contains at least one NA value among n observations. Without loss of generality, let $X_i = (X_{i1}, \ldots, X_{iq}, X_{i,q+1}, \ldots, X_{ip})$, where each of the first q variables contains at least one NA value among n observations and the remaining $p - q$ variables have no NA value. Define

$$I_{ij} = I(X_{ij} = \texttt{NA}), i = 1, \ldots, n; j = 1, \ldots, q.$$

Then, for each covariate $(X_{1j}, \ldots, X_{nj})^{\mathrm{T}}$ with NA value, $j = 1, \ldots, q$, we can replace NA value by any number (say, the mean of the observed values of covariate j if it is continuous or the majority level of covariate j if it is categorical). Thus, let

$$W_i = (X_{i1}, I_{i1}, X_{i2}, I_{i2}, \ldots, X_{iq}, I_{iq}, X_{i,q+1}, \ldots, X_{ip})$$

be the expanded vector of covariates. Thus, the observed dataset becomes $\{O_i = (W_i, A_i, Y_i), i = 1, \ldots, n\}$. Thanks to super learner, high-dimensionality of W is not an issue.

8.5.3 How to use TMLE for RCTs?

The main difference between an interventional study and a non-interventional study is that the propensity score function is known in the interventional study. For example, in an RCT using complete randomization with ratio r:1 between treatments $Z = 1$ and $Z = 0$, the propensity score function is $\mathbb{P}(Z = 1) = r/(1+r)$ for each subject. In a stratified RCT, if the stratification factor is S, with $S = 1, \ldots, K$, and the randomization ratio is r_k:1 within stratum $S = k$, then the propensity score function is $\mathbb{P}(Z = 1|S = k) = r_k/(1 + r_k)$.

In the application of TMLE to RCTs, we can consider a saturate logistic regression model, $Z \sim 1$ for complete randomization and $Z \sim S$ for stratified randomization, respectively, to model the propensity score function g. Thanks to randomization, this model for g is correctly specified for sure. At the same time, we consider super learner to model the outcome regression function Q. Since TMLE is doubly robust in the sense that it is consistent if either g is consistently estimated or Q is consistently estimated, we can conclude that TMLE is consistent for sure because g is known or can be estimated by a saturate logistic regression model consistently for sure.

8.5.4 How to implement TMLE?

We can implement TMLE using R package "tmle" (Gruber and van der Laan 2012). Some of the key arguments of R function "tmle" are shown in what follows and refer to Gruber and van der Laan (2012) for its full syntax.

```
tmle(Y, A, W, Delta = rep(1,length(Y)),
    Qform = NULL, Q.SL.library = c("SL.glm", "tmle.SL.dbarts2", "SL.glmnet"),
    g1W = NULL, gform = NULL, g.SL.library = c("SL.glm", "tmle.SL.dbarts.k.5", "SL.gam"),
    g.Deltaform = NULL, g.Delta.SL.library = c("SL.glm", "tmle.SL.dbarts.k.5", "SL.gam"),
    family = "gaussian", ...)
```

Here are some explanations of the above key arguments:

- `Y`: column name of the outcome variable;
- `A`: column name of the treatment variable;
- `W`: column names of the covariates (or expanded vector of covariates including the NA value indicators if any);
- `Delta`: column name of missing outcome indicator (1 means observed; 0 means missing). Note: This is different from Δ used in this book, with $\Delta = 1$ for missing and $\Delta = 0$ for observed;
- `Qform`: optional regression formula of form $Y \sim A + W$ for estimation of outcome regression function Q, suitable for call to `glm`;
- `Q.SL.library`: optional vector of predictive algorithms to use for `SuperLearner` estimation of initial Q;
- `g1W`: optional vector of conditional treatment assignment probabilities for RCTs in which the treatment assignment probabilities are known;
- `gform`: optional regression formula of the form $A \sim W$ for estimation of propensity score function g, if specified this overrides the call to `SuperLearner`;
- `g.SL.library`: optional vector of predictive algorithms to use for `SuperLearner` estimation of initial g;
- `g.Deltaform`: optional regression formula of the form `Delta` $\sim A + W$ for estimation of the non-missing probability function, if specified this overrides the call to `SuperLearner`;

- `g.SL.library`: optional vector of predictive algorithms to use for `SuperLearner` estimation of the non-missing probability function;
- `family`: specification for working regression models, generally 'gaussian' for continuous outcomes (default), 'binomial' for binary outcomes.

The outputs of R function "tmle" include estimates for both ATE estimand and ATT estimand, their standard errors, confidence interval estimates, and the corresponding p-values.

8.6 Exercises

A real dataset of the NHEFS study (National Health and Nutrition Examination Survey Data I Epidemiologic Follow-up Study) was prepared by Hernán and Robins for their book (Hernán and Robins 2020). Hernán and Robins (2020) used the dataset throughout their book and stated the following statement: "encourage readers to improve upon and refine our analyses." The dataset can be downloaded from their book website, https://www.hsph.harvard.edu/miguel-hernan/causal-inference-book.

Ex 8.1

Download the above NHEFS dataset in a CSV format. Consider the following three steps to summarize 11 variables.

1. Summarize 9 covariates and denote them as W:
   (a) sex (0: male, 1: female)
   (b) race (0: white, 1: other)
   (c) age
   (d) education (1: 8th grade or less, 2: high school dropout, 3: high school, 4: college dropout, 5: college or more) cigarettes per day
   (e) years of smoking
   (f) exercise (0: much exercise, 1: moderate exercise, 2: little or no exercise)
   (g) active (0: very active, 1: moderately active, 2: inactive)
   (h) weight in 1971 in kg.
2. Summarize variable named "qsmk" and denote it as A.
3. Summarize variable named "wt82_71" and denote it as Y.

Ex 8.2

Use R function "tmle" to obtain an AIPW estimate for statistical estimand θ_{ATE}, fitting a linear model for outcome regression function Q and a logistic model for propensity score function g.

Ex 8.3

Use R function "tmle" to obtain a TMLE estimate for statistical estimand θ_{ATE}, using the default SL library for Q and g, respectively.

Ex 8.4

Use R function "tmle" to obtain a TMLE estimate for statistical estimand θ_{ATE}, using the default SL library for Q and g, respectively.

9

The Art of Estimation (III): LTMLE

9.1 Longitudinal Cohort Studies

In Chapter 8, we described the targeted learning framework (van der Laan and Rose 2011) for cohort studies; in particular, we reviewed the targeted maximum likelihood estimator or targeted minimum loss estimator (TMLE). In this chapter, we will describe the longitudinal TMLE (LTMLE) for longitudinal cohort studies discussed in Chapter 6.

9.1.1 Causal estimand

As in Chapter 6, let $\overline{A}(t) = (A(0), \ldots, A(t))$ be the observed treatment sequence up to t, and let $\overline{X}(t) = (X(0), \ldots, X(t))$ be the vector consisting of all the observed history up to time t including baseline covariates, time-dependent covariates, and intermediate outcomes, $t = 0, \ldots, T-1$. In particular, define $\overline{A} = (A(0), \ldots, A(T-1))$ and $\overline{X} = (X(0), \ldots, X(T-1))$. Let Y be the primary outcome variable measured at time T.

Let $Y^{\overline{a}}$ be the potential outcome had the patient been treated by treatment sequence $\overline{a} = (a(0), a(1), \ldots, a(T-1))$. Let $\overline{a}(t) = (a(0), a(1), \ldots, a(t))$. Two particular treatment sequences are $\overline{a} = \overline{1} = (1, \ldots, 1)$ and $\overline{a} = \overline{0} = (0, \ldots, 0)$. After defining the above potential outcomes, we can define the causal estimand of interest. In this chapter, we focus on the following estimand,

$$v^*(\overline{a}) = \mathbb{E}(Y^{\overline{a}}), \tag{9.1}$$

which is referred to as the *value* of $\overline{a}$ (Tsiatis et al. 2020).

DOI: 10.1201/9781003433378-9

The reason why we focus on the above estimand is that many other estimands can be defined consequently; e.g., the average treatment effect (ATE) of treatment sequence $\overline{1}$ compared with treatment sequence $\overline{0}$,

$$\theta^*_{\text{ATE}} = \mathbb{E}(Y^{\overline{a}=\overline{1}}) - \mathbb{E}(Y^{\overline{a}=\overline{0}}) = v^*(\overline{1}) - v^*(\overline{0}). \tag{9.2}$$

9.1.2 Identification

In Chapter 6, we showed that, under the identifiability assumptions (consistency, sequential exchangeability, and positivity), we can translate the ATE causal estimand (9.2) into a statistical estimand, by either the standardization strategy or the weighting strategy. In this chapter, we revisit the task of identification, but for the value causal estimand (9.1), via the weighting strategy and a series of standardization strategies.

The weighting strategy

In Chapter 6, we showed the following result for the setting where $T = 2$,

$$\begin{aligned} v^*(\overline{a}) &= \mathbb{E}(Y^{\overline{a}}) \\ &= \mathbb{E}\left[\frac{I(\overline{A} = \overline{a})Y}{\mathbb{P}\{A(0) = a(0)|X(0)\}\mathbb{P}\{A(1) = a(1)|X(0), A(0), X(1)\}}\right] \\ &\triangleq v(\overline{a}), \end{aligned}$$

under the identifiability assumptions. We can generate it to any $T > 1$,

$$\begin{aligned} v(\overline{a}) =& \mathbb{E}\left[\frac{I(\overline{A} = \overline{a})Y}{\mathbb{P}\{A(0) = a(0)|X(0)\}\prod_{t=1}^{T-1}\mathbb{P}\{A(t) = a(t)|\overline{X}(t), \overline{A}(t-1)\}}\right] \\ =& \mathbb{E}\left[\frac{I(\overline{A} = \overline{a})Y}{\prod_{t=0}^{T-1}\mathbb{P}\{A(t) = a(t)|\overline{X}(t), \overline{A}(t-1)\}}\right], \end{aligned} \tag{9.3}$$

where, by convention, $\overline{X}(0) = X(0)$, $\overline{A}(0) = A(0)$, and $\overline{X}(-1) = \overline{A}(-1) = \emptyset$.

Define the following propensity score function at time t,

$$g_t\left(a(t)|\overline{X}(t), \overline{A}(t-1)\right) = \mathbb{P}\left\{A(t) = a(t)|\overline{X}(t), \overline{A}(t-1)\right\}, \tag{9.4}$$

where $t = 0, \ldots, T-1$. Then, (9.3) can be expressed as

$$v(\overline{a}) = \mathbb{E}\left[\frac{I(\overline{A} = \overline{a})Y}{\prod_{t=0}^{T-1} g_t\left(a(t)|\overline{X}(t), \overline{A}(t-1)\right)}\right]. \tag{9.5}$$

A series of standardization strategies

We consider a backward procedure. Starting from (9.5), we have

$$
\begin{aligned}
& v(\overline{a}) \\
& \overset{(a)}{=} \mathbb{E}\left\{\mathbb{E}\left[\frac{I(\overline{A}=\overline{a})Y}{\prod_{t=0}^{T-1} g_t\left(a(t)|\overline{X}(t),\overline{A}(t-1)\right)}\middle|\overline{X}(T-1),\overline{A}(T-2)\right]\right\} \\
& \overset{(b)}{=} \mathbb{E}\left\{\mathbb{E}\left[\frac{I(\overline{A}=\overline{a})Y^{\overline{a}}}{\prod_{t=0}^{T-1} g_t\left(a(t)|\overline{X}(t),\overline{A}(t-1)\right)}\middle|\overline{X}(T-1),\overline{A}(T-2)\right]\right\} \\
& \overset{(c)}{=} \mathbb{E}\left\{\frac{I\left(\overline{A}(T-2)=\overline{a}(T-2)\right)}{\prod_{t=0}^{T-2} g_t\left(a(t)|\overline{X}(t),\overline{A}(t-1)\right)}\right. \\
& \qquad \left.\times \mathbb{E}\left[\frac{I\left(\overline{A}(T-1)=\overline{a}(T-1)\right)Y^{\overline{a}}}{g_{T-1}\left(a(T-1)|\overline{X}(T-1),\overline{A}(T-2)\right)}\middle|\overline{X}(T-1),\overline{A}(T-2)\right]\right\} \\
& \overset{(d)}{=} \mathbb{E}\left\{\frac{I\left(\overline{A}(T-2)=\overline{a}(T-2)\right)}{\prod_{t=0}^{T-2} g_t\left(a(t)|\overline{X}(t),\overline{A}(t-1)\right)}\mathbb{E}\left[Y^{\overline{a}}|\overline{X}(T-1),\overline{A}(T-2)\right]\right\} \\
& \overset{(e)}{=} \mathbb{E}\left\{\frac{I\left(\overline{A}(T-2)=\overline{a}(T-2)\right)}{\prod_{t=0}^{T-2} g_t\left(a(t)|\overline{X}(t),\overline{A}(t-1)\right)}\right. \\
& \qquad \left.\times \mathbb{E}\left[Y^{\overline{a}}|\overline{X}(T-1),\overline{A}(T-2),A(T-1)=a(T-1)\right]\right\} \\
& \overset{(f)}{=} \mathbb{E}\left\{\mathbb{E}\left[\frac{I\left(\overline{A}(T-2)=\overline{a}(T-2)\right)Y^{\overline{a}}}{\prod_{t=0}^{T-2} g_t\left(a(t)|\overline{X}(t),\overline{A}(t-1)\right)}\middle|\overline{X}(T-1),\overline{A}(T-2),a(T-1)\right]\right\} \\
& \overset{(g)}{=} \mathbb{E}\left\{\mathbb{E}\left[\frac{I\left(\overline{A}(T-2)=\overline{a}(T-2)\right)Y}{\prod_{t=0}^{T-2} g_t\left(a(t)|\overline{X}(t),\overline{A}(t-1)\right)}\middle|\overline{X}(T-1),\overline{A}(T-2),a(T-1)\right]\right\} \\
& \overset{(h)}{=} \mathbb{E}\left\{\frac{I\left(\overline{A}(T-2)=\overline{a}(T-2)\right)}{\prod_{t=0}^{T-2} g_t\left(a(t)|\overline{X}(t),\overline{A}(t-1)\right)}\mathbb{E}\left[Y|\overline{X}(T-1),\overline{A}(T-2),a(T-1)\right]\right\} \\
& \overset{(i)}{=} \mathbb{E}\left\{\frac{I\left(\overline{A}(T-2)=\overline{a}(T-2)\right)}{\prod_{t=0}^{T-2} g_t\left(a(t)|\overline{X}(t),\overline{A}(t-1)\right)}Q_{T-1}\left(\overline{X}(T-1),\overline{A}(T-2),a(T-1)\right)\right\},
\end{aligned}
$$

where (a) is using the law of iterated expectations, (b) holds under the consistency assumption, (c) holds because of the conditional expectation, (d) holds because g_{T-1} is canceled out, (e) holds under the sequential exchangeability assumption, (f) holds because of the conditional expectation, (g) holds under the consistency assumption, (h) holds because of the conditional expectation, and in (i)

$$Q_{T-1}\left(\overline{X}(T-1),\overline{A}(T-1)\right)=\mathbb{E}\left[Y|\overline{X}(T-1),\overline{A}(T-1)\right] \tag{9.6}$$

is the outcome regression function of $Y \sim \overline{X}(T-1)+\overline{A}(T-1)$.

Next, we can proceed one step backward. In the previous step the outcome is Y. In this step, consider the following tentative "outcome."

$$\widetilde{Y}_{T-1} = Q_{T-1}\left(\overline{X}(T-1), \overline{A}(T-2), a(T-1)\right),$$

which equals the regression function evaluated at $\overline{X}(T-1), \overline{A}(T-2)$, and $A(T-1) = a(T-1)$. Then following the similar arguments, we can show that

$$\begin{aligned}
&v(\overline{a})\\
=&\mathbb{E}\left\{\frac{I\left(\overline{A}(T-2)=\overline{a}(T-2)\right)}{\prod_{t=0}^{T-2} g_{t0}\left(a(t)|\overline{X}(t), \overline{A}(t-1)\right)} Q_{T-1}\left(\overline{X}(T-1), \overline{A}(T-2), a(T-1)\right)\right\}\\
=&\mathbb{E}\left\{\frac{I\left(\overline{A}(T-2)=\overline{a}(T-2)\right)}{\prod_{t=0}^{T-2} g_{t0}\left(a(t)|\overline{X}(t), \overline{A}(t-1)\right)} \widetilde{Y}_{T-1}\right\}\\
=&\mathbb{E}\left\{\frac{I\left(\overline{A}(T-3)=\overline{a}(T-3)\right)}{\prod_{t=0}^{T-3} g_{t0}\left(a(t)|\overline{X}(t), \overline{A}(t-1)\right)} Q_{T-2}\left(\overline{X}(T-2), \overline{A}(T-3), a(T-2)\right)\right\},
\end{aligned}$$

where

$$Q_{T-2}\left(\overline{X}(T-2), \overline{A}(T-2)\right) = \mathbb{E}\left[\widetilde{Y}_{T-1}\middle|\overline{X}(T-2), \overline{A}(T-2)\right] \tag{9.7}$$

is the outcome regression function of $\widetilde{Y}_{T-1} \sim \overline{X}(T-2) + \overline{A}(T-2)$.

We can repeat the above by $T-2$ steps backward until we reach the last step. In the last step, consider the following tentative "outcome,"

$$\widetilde{Y}_1 = Q_1\left(\overline{X}(1), A(0), a(1)\right),$$

which equals the regression function $Q_1(\overline{X}(1), \overline{A}(1))$ evaluated at $\overline{X}(T-1), \overline{A}(T-2)$, and $A(T-1) = a(T-1)$. Similarly, we can show that

$$\begin{aligned}
&v(\overline{a})\\
=&\mathbb{E}\left\{\frac{I\left(A(0)=a(0)\right)}{g_0\left(a(0)|X(0)\right)} Q_1\left(\overline{X}(1), A(0), a(1)\right)\right\}\\
=&\mathbb{E}\left\{\frac{I\left(A(0)=a(0)\right)}{g_0\left(a(0)|X(0)\right)} \widetilde{Y}_1\right\}\\
=&\mathbb{E}\left\{Q_0\left(X(0), a(0)\right)\right\},
\end{aligned}$$

where

$$Q_0\left(X(0), A(0)\right) = \mathbb{E}\left[\widetilde{Y}_1\middle|X(0), A(0)\right] \tag{9.8}$$

is the outcome regression function of $\widetilde{Y}_1 \sim X(0) + A(0)$. The last step is similar to applying the standardization strategy to the study.

9.1.3 Efficient influence function

We will use the shortcut proposed in Chapter 8 to find the efficient influence function for estimating $v(\overline{a})$. For this aim, we center all the above T versions of the statistical estimand. This time, let's take these steps forward.

At Step $t = 0$, for the following version of statistical estimand,

$$v(\overline{a}) = \mathbb{E}\left\{Q_0\left(X(0), a(0)\right)\right\}, \tag{9.9}$$

we define

$$\phi_0(O) = Q_0\left(X(0), a(0)\right) - v(\overline{a}).$$

At Step $t = 2, \ldots, T-1$, for the following version of statistical estimand,

$$v(\overline{a}) = \mathbb{E}\left\{\frac{I\left(\overline{A}(t-1) = \overline{a}(t-1)\right)}{\prod_{s=0}^{t-1} g_s\left(a(s)|\overline{X}(s), \overline{A}(s-1)\right)} Q_t\left(\overline{X}(t), \overline{A}(t-1), a(t)\right)\right\}, \tag{9.10}$$

we define

$$\begin{aligned}\phi_t(O) = &\frac{I\left(\overline{A}(t-1) = \overline{a}(t-1)\right)}{\prod_{s=0}^{t-1} g_s\left(a(s)|\overline{X}(s), \overline{A}(s-1)\right)} \\ &\times \left[Q_t\left(\overline{X}(t), \overline{A}(t-1), a(t)\right) - Q_{t-1}\left(\overline{X}(t-1), \overline{A}(t-1)\right)\right].\end{aligned}$$

At Step $t = T$, for the following version of statistical estimand,

$$v(\overline{a}) = \mathbb{E}\left[\frac{I(\overline{A} = \overline{a})Y}{\prod_{t=0}^{T-1} g_t\left(a(t)|\overline{X}(t), \overline{A}(t-1)\right)}\right], \tag{9.11}$$

we define

$$\phi_T(O) = \frac{I(\overline{A} = \overline{a})}{\prod_{t=0}^{T-1} g_t\left(a(t)|\overline{X}(t), \overline{A}(t-1)\right)}\left[Y - Q_{T-1}\left(\overline{X}(T-1), \overline{A}(T-1)\right)\right].$$

By the shortcut of finding the efficient influence function, we can show that the efficient influence function of $v(\overline{a})$ is

$$\phi_{\text{EIF-}\overline{a}}(O) = \sum_{t=0}^{T} \phi_t(O). \tag{9.12}$$

That is,

$$\begin{aligned}\phi_{\text{EIF-}\overline{a}}(O) = \left[Q_0\left(X(0), a(0)\right) - v(\overline{a})\right] + &\sum_{t=1}^{T-1} \frac{I\left(\overline{A}(t-1) = \overline{a}(t-1)\right)}{\prod_{s=0}^{t-1} g_s\left(a(s)|\overline{X}(s), \overline{A}(s-1)\right)} \\ &\times \left[Q_t\left(\overline{X}(t), \overline{A}(t-1), a(t)\right) - Q_{t-1}\left(\overline{X}(t-1), \overline{A}(t-1)\right)\right] \\ + &\frac{I(\overline{A} = \overline{a})}{\prod_{t=0}^{T-1} g_t\left(a(t)|\overline{X}(t), \overline{A}(t-1)\right)}\left[Y - Q_{T-1}\left(\overline{X}(T-1), \overline{A}(T-1)\right)\right].\end{aligned}$$

Proof of the validity of the shortcut:

Consider any regular parametric submodel indicated by $\boldsymbol{\epsilon} = (\epsilon^X, \epsilon_t^g, \epsilon_t^Q, t = 0, \ldots, T-1)^{\mathrm{T}}$, where ϵ^X is for $f_X(x)$, ϵ_t^g is for g_t, and ϵ_t^Q is for Q_t, $t = 0, \ldots, T-1$. Assume that when $\boldsymbol{\epsilon} = \mathbf{0} = (\mathbf{0}, \ldots, \mathbf{0})^{\mathrm{T}}$, these parametric submodels are the same as the true underlying models. Let $v(\overline{a}; \boldsymbol{\epsilon})$ be the statistical estimand which is defined via any of the above T versions, which is a function of $\boldsymbol{\epsilon}$ had the data been generated by the parametric submodel.

We can verify that function $\phi_{\text{EFF-}\overline{a}}(O)$ defined via the shortcut satisfies the condition of the fundamental theorem regularity provided in Chapter 8,

$$\partial v(\overline{a}; \boldsymbol{\epsilon})/\partial \boldsymbol{\epsilon}|_{\boldsymbol{\epsilon}=\mathbf{0}} = \mathbb{E}[\phi_{\text{EFF-}\overline{a}}(O) S_{\boldsymbol{\epsilon}}],$$

where $S_{\boldsymbol{\epsilon}} = (S_{\epsilon^X}, S_{\epsilon_0^g}, S_{\epsilon_0^Q}, \ldots, S_{\epsilon_{T-1}^g}, S_{\epsilon_{T-1}^Q})^{\mathrm{T}}$ is the corresponding score.

For ϵ^X, we use version (9.9) to define $v(\overline{a}; \boldsymbol{\epsilon})$. Thus, we have

$$\begin{aligned} \partial v(\overline{a}; \boldsymbol{\epsilon})/\partial \epsilon^X|_{\boldsymbol{\epsilon}=\mathbf{0}} &= \mathbb{E}[\phi_0(O) S_{\epsilon^X}], \\ \mathbb{E}[\phi_s(O) S_{\epsilon^X}] &= 0, s = 1, \ldots, T. \end{aligned}$$

This implies that $\partial v(\overline{a}; \boldsymbol{\epsilon})/\partial \epsilon^X|_{\boldsymbol{\epsilon}=\mathbf{0}} = \mathbb{E}[\phi_{\text{EFF-}\overline{a}}(O) S_{\epsilon^X}]$.

For ϵ_0^g, we also use version (9.9) to define $v(\overline{a}; \boldsymbol{\epsilon})$. Thus,

$$\begin{aligned} \partial v(\overline{a}; \boldsymbol{\epsilon})/\partial \epsilon_0^g|_{\boldsymbol{\epsilon}=\mathbf{0}} &= \mathbb{E}[\phi_0(O) S_{\epsilon_0^g}], \\ \mathbb{E}[\phi_s(O) S_{\epsilon_0^g}] &= 0, s = 1, \ldots, T. \end{aligned}$$

This implies that $\partial v(\overline{a}; \boldsymbol{\epsilon})/\partial \epsilon_0^g|_{\boldsymbol{\epsilon}=\mathbf{0}} = \mathbb{E}[\phi_{\text{EFF-}\overline{a}}(O) S_{\epsilon_0^g}]$.

For $\epsilon_t^g, t \geq 1$, we use version (9.10) to define $v(\overline{a}; \boldsymbol{\epsilon})$. Thus,

$$\begin{aligned} \partial v(\overline{a}; \boldsymbol{\epsilon})/\partial \epsilon_t^g|_{\boldsymbol{\epsilon}=\mathbf{0}} &= \mathbb{E}[\phi_t(O) S_{\epsilon_t^g}], \\ \mathbb{E}[\phi_s(O) S_{\epsilon_t^g}] &= 0, s = 0, \ldots, T \text{ and } s \neq t. \end{aligned}$$

This implies that $\partial v(\overline{a}; \boldsymbol{\epsilon})/\partial \epsilon_t^g|_{\boldsymbol{\epsilon}=\mathbf{0}} = \mathbb{E}[\phi_{\text{EFF-}\overline{a}}(O) S_{\epsilon_t^g}]$, $t = 1, \ldots, T-1$.

For $\epsilon_t^Q, t \geq 0$, we use version (9.11) to define $v(\overline{a}; \boldsymbol{\epsilon})$. Thus,

$$\begin{aligned} \partial v(\overline{a}; \boldsymbol{\epsilon})/\partial \epsilon_t^Q|_{\boldsymbol{\epsilon}=\mathbf{0}} &= \mathbb{E}[\phi_T(O) S_{\epsilon_t^Q}], \\ \mathbb{E}[\phi_s(O) S_{\epsilon_t^Q}] &= 0, s = 0, \ldots, T-1. \end{aligned}$$

This implies that $\partial v(\overline{a}; \boldsymbol{\epsilon})/\partial \epsilon_t^Q|_{\boldsymbol{\epsilon}=\mathbf{0}} = \mathbb{E}[\phi_{\text{EFF-}\overline{a}}(O) S_{\epsilon_t^Q}]$, $t = 0, \ldots, T-1$.

Therefore, we show that $\partial v(\overline{a}; \boldsymbol{\epsilon})/\partial \boldsymbol{\epsilon}|_{\boldsymbol{\epsilon}=\mathbf{0}} = \mathbb{E}[\phi_{\text{EFF-}\overline{a}}(O) S_{\boldsymbol{\epsilon}}]$. □

9.1.4 LTMLE

After we obtain the efficient influence function, $\phi_{\text{EFF-}\overline{a}}(O)$, we are ready to describe the LTMLE procedure. Simply put, LTMLE is a combination of a series of TMLE procedures that are applied backward.

Using super learner, we can obtain initial estimators of g_t and Q_t, $t = 0, \ldots, T-1$. The first step is to obtain the targeted estimator of Q_{T-1}, denoted as $\widehat{Q}^*_{T-1}$, from the regression modeling of $Y \sim (\overline{X}, \overline{A})$, using TMLE with the following clever covariate,

$$H_{T-1}\left(\overline{X}, \overline{A}\right) = \frac{I(\overline{A} = \overline{a})}{\prod_{t=0}^{T-1} g_t\left(A(t)|\overline{X}(t), \overline{A}(t-1)\right)}.$$

In the next step, denote the tentative "outcome" as

$$\widetilde{Y}_{T-1} = \widehat{Q}^*_{T-1}\left(\overline{X}(T-1), \overline{A}(T-2), a(T-1)\right).$$

Then obtain the targeted estimator of Q_{T-2}, denoted as $\widehat{Q}^*_{T-2}$, from the regression modeling of $\widetilde{Y}_{T-1} \sim (\overline{X}(T-2), \overline{A}(T-2))$, using TMLE with the following clever covariate,

$$H_{T-2}\left(\overline{X}(T-2), \overline{A}(T-2)\right) = \frac{I\left(\overline{A}(T-2) = \overline{a}(T-2)\right)}{\prod_{t=0}^{T-2} g_t\left(A(t)|\overline{X}(t), \overline{A}(t-1)\right)}.$$

Repeat the above TMLE procedures until the last step. At the last step, denote the tentative "outcome" as

$$\widetilde{Y}_1 = \widehat{Q}^*_1\left(\overline{X}(1), A(0), a(1)\right).$$

Then obtain the targeted estimator of Q_0, denoted as $\widehat{Q}^*_0$, from the regression modeling of $\widetilde{Y}_1 \sim (X(0), A(0))$, using TMLE with the following clever covariate,

$$H_0\left(X(0), A(0)\right) = \frac{I\left(\overline{A}(0) = \overline{a}(0)\right)}{g_0\left(A(0)|X(0)\right)}.$$

Finally, we obtain the LTMLE estimator,

$$\widehat{v}(\overline{a})_{\text{LTMLE}} = \frac{1}{n}\sum_{i=1}^{n} \widehat{Q}^*_0\left(X_i(0), a(0)\right). \tag{9.13}$$

9.1.5 ATE estimand

After we discuss the efficient influence function for the value estimand, we can obtain the efficient influence function for the ATE estimand easily. Under the identifiability assumptions, we can translate causal estimand θ^*_{ATE} in the following statistical estimand,

$$\theta_{\text{ATE}} = v(\overline{1}) - v(\overline{0}).$$

We can easily verify that the efficient influence function for estimating θ_{ATE} is

$$\phi_{\text{EIF-ATE}}(O) = \phi_{\text{EFF-}\overline{1}}(O) - \phi_{\text{EFF-}\overline{0}}(O). \tag{9.14}$$

Thus, the LTMLE estimator of θ_{ATE} is

$$\widehat{\theta}_{\text{LTMLE-ATE}} = \widehat{v}(\overline{1})_{\text{LTMLE}} - \widehat{v}(\overline{0})_{\text{LTMLE}}. \tag{9.15}$$

And the asymptotic variance of $\sqrt{n}(\widehat{\theta}_{\text{LTMLE-ATE}} - \theta_{\text{ATE}})$ is $\mathbb{V}(\phi_{\text{EIF-ATE}}(O))$, which can be estimated by its sample analog.

9.2 Missing Data

9.2.1 Monotone missing

In this subsection, we only consider monotone missing data due to analysis dropouts. Later we will discuss how to handle missing data and intercurrent events (ICEs) in general.

Let $\Delta(t)$ be the indicator of analysis dropout at time t, $t = 1, \ldots, T$. Let $\overline{\Delta} = (\Delta(1), \ldots, \Delta(T))$ and $\overline{\Delta}(t) = (\Delta(1), \ldots, \Delta(t))$, $t = 1, \ldots, T$. In this subsection, we assume that the analysis dropout pattern is monotone; that is, if $\Delta(t) = 1$, then $\Delta(t') = 1$ for $t' = t+1, \ldots, T$. Therefore, we can understand such monotone missing as censoring.

Estimand

Let $Y^{\overline{a}, \overline{\delta}=\overline{0}}$ be the potential outcome if the patient had followed treatment sequence $\overline{a}$ and the data had been collected throughout the study.

After defining the potential outcome, we can define the causal estimand of interest. We start with the following estimand,

$$v_{\text{H}}^{*}(\overline{a}) = \mathbb{E}(Y^{\overline{a}, \overline{\delta}=\overline{0}}). \tag{9.16}$$

Consequently, many other estimands can be defined; for example,

$$\theta_{\text{H}}^{*} = \mathbb{E}(Y^{\overline{a}=\overline{1}, \overline{\delta}=\overline{0}}) - \mathbb{E}(Y^{\overline{a}=\overline{0}, \overline{\delta}=\overline{0}}) = v_{\text{H}}^{*}(\overline{1}) - v_{\text{H}}^{*}(\overline{0}). \tag{9.17}$$

Identification

Besides those three identifiability assumptions (consistency, sequential exchangeability, and positivity), we assume the missing at random assumption.

If we think of "$\overline{A} = \overline{a}, \overline{\Delta} = 0$" as an action parallel to action "$\overline{A} = \overline{a}$," we can revise the statements in the preceding section to accomplish the task of identification. For this aim, define the following non-missing (or non-censoring) probability function,

$$h_t\left(\overline{X}(t), \overline{A}(t)\right) = \mathbb{P}\left\{\Delta(t+1) = 0 \middle| \overline{X}(t), \overline{A}(t), \overline{\Delta}(t) = \overline{0}\right\},$$

where $t = 0, \ldots, T-1$ and by convention, $\overline{\Delta}(0) = \emptyset$.

To simply the notation, let

$$\bar{g}_t = \prod_{s=0}^{t} g_s\left(A(s)|\overline{X}(s), \overline{A}(s-1)\right),$$
$$\bar{h}_t = \prod_{s=0}^{t} h_s\left(\overline{X}(s), \overline{A}(s)\right),$$
$$\bar{I}_t = I(\overline{A}(t) = \bar{a}(t), \overline{\Delta}(t+1) = \bar{0}),$$

where $t = 0, \ldots, T-1$.

Similarly, the statistical estimand to be developed has T equivalent forms. First, we obtain the following form via the weighting strategy,

$$v_{\mathrm{H}}(\bar{a}) = \mathbb{E}\left[\frac{\bar{I}_{T-1}Y}{\bar{g}_{T-1}\bar{h}_{T-1}}\right].$$

Second, consider Y as the outcome variable and define

$$\widetilde{Q}_{T-1}\left(\overline{X}(T-1), \overline{A}(T-1)\right) = \mathbb{E}\left\{Y|\overline{X}(T-1), \overline{A}(T-1), \overline{\Delta} = \bar{0}\right\}$$

as the outcome regression function in the modeling of $Y \sim (\overline{X}(T-1), \overline{A}(T-1))$ conditional on $\overline{\Delta} = \bar{0}$. Thus, we obtain the second form,

$$v_{\mathrm{H}}(\bar{a}) = \mathbb{E}\left[\frac{\bar{I}_{T-2}}{\bar{g}_{T-2}\bar{h}_{T-2}}\widetilde{Q}_{T-1}\left(\overline{X}(T-1), \overline{A}(T-2), a(T-1)\right)\right].$$

Third, consider the following variable as the tentative "outcome,"

$$\widetilde{Y}^{*}_{T-1} = \widetilde{Q}_{T-1}\left(\overline{X}(T-1), \overline{A}(T-2), a(T-1)\right),$$

and define

$$\widetilde{Q}_{T-2}\left(\overline{X}(T-2), \overline{A}(T-2)\right) = \mathbb{E}\left\{\widetilde{Y}^{*}_{T-1}|\overline{X}(T-2), \overline{A}(T-2), \overline{\Delta}(T-1) = \bar{0}\right\}$$

as the outcome regression function in the modeling of $\widetilde{Y}^{*}_{T-1} \sim (\overline{X}(T-2), \overline{A}(T-2))$ conditional on $\overline{\Delta}(T-1) = \bar{0}$. Thus, we obtain the third form,

$$v_{\mathrm{H}}(\bar{a}) = \mathbb{E}\left[\frac{\bar{I}_{T-3}}{\bar{g}_{T-3}\bar{h}_{T-3}}\widetilde{Q}_{T-2}\left(\overline{X}(T-2), \overline{A}(T-3), a(T-2)\right)\right].$$

Repeat the above steps and obtain the following forms,

$$v_{\mathrm{H}}(\bar{a}) = \mathbb{E}\left[\frac{\bar{I}_{t-1}}{\bar{g}_{t-1}\bar{h}_{t-1}}\widetilde{Q}_{t}\left(\overline{X}(t), \overline{A}(t-1), a(t)\right)\right],$$

where $t = T-1, T-2, \ldots, 0$. This includes the last form of the estimand,

$$v_{\mathrm{H}}(\bar{a}) = \mathbb{E}\left[\widetilde{Q}_{0}\left(X(0), a(0)\right)\right].$$

Efficient influence function

By the shortcut of finding the efficient influence function, the efficient influence function in estimating $v_{\mathrm{H}}(\overline{a})$ is

$$\begin{aligned}
&\phi_{\text{EIF-H-}\overline{a}}(O)\\
=&\left[\widetilde{Q}_0\left(X(0), a(0)\right) - v(\overline{a})\right]\\
&+\sum_{t=1}^{T-1} \frac{\overline{I}_{t-1}}{\overline{g}_{t-1}\overline{h}_{t-1}} \left[\widetilde{Q}_t\left(\overline{X}(t), \overline{A}(t-1), a(t)\right) - \widetilde{Q}_{t-1}\left(\overline{X}(t-1), \overline{A}(t-1)\right)\right]\\
&+\frac{\overline{I}_{T-1}}{\overline{g}_{T-1}\overline{h}_{T-1}} \left[Y - \widetilde{Q}_{T-1}\left(\overline{X}(T-1), \overline{A}(T-1)\right)\right].
\end{aligned} \tag{9.18}$$

LTMLE

After we derive the efficient influence function, $\phi_{\text{EFF-H-}\overline{a}}(O)$, we are ready to describe the LTMLE procedure, which is a combination of a series of TMLE procedures that are applied backward. Here we omit the details because the LTME procedure is the same as the one described in the preceding section except that now clever covariate in each TMLE procedure becomes

$$\begin{aligned}
\widetilde{H}_t\left(\overline{X}(t), \overline{A}(t)\right) =& \frac{\overline{I}_{t-1}}{\overline{g}_{t-1}\overline{h}_{t-1}}\\
=& \frac{I(\overline{A}(t) = \overline{a}(t)) \cdot I(\overline{\Delta}(t+1) = \overline{0})}{\prod_{s=0}^{t-1} g_s\left(A(s)|\overline{X}(s), \overline{A}(s-1)\right) \cdot \prod_{s=0}^{t-1} h_s\left(\overline{X}(s), \overline{A}(s)\right)},
\end{aligned}$$

where $t = 0, \ldots, T-1$.

ATE estimand

After we derive the efficient influence function for the value estimand, we can derive the efficient influence function for the ATE estimand easily. Under the identifiability assumptions, we can translate the causal estimand θ^*_{ATE} into the following statistical estimand,

$$\theta_{\mathrm{H}} = v(\overline{1}) - v(\overline{0}).$$

We can easily verify that the efficient influence function for estimating θ_{H} is

$$\phi_{\text{EIF-H}}(O) = \phi_{\text{EFF-H-}\overline{1}}(O) - \phi_{\text{EFF-H-}\overline{0}}(O). \tag{9.19}$$

Thus, the LTLME estimator of θ_{H} is

$$\widehat{\theta}_{\text{LTMLE-H}} = \widehat{v}(\overline{1})_{\text{LTMLE-H}} - \widehat{v}(\overline{0})_{\text{LTMLE-H}}.$$

And the asymptotic variance of $\sqrt{n}(\widehat{\theta}_{\text{LTMLE-H}} - \theta_{\mathrm{H}})$ is $\mathbb{V}(\phi_{\text{EIF-H}}(O))$, which can be estimated by its sample analog.

9.2.2 Non-monotone missing

More complicated than monotone missing or censoring, missing may be non-monotone. That is, there may be missing data or non-available (NA) values in some baseline covariates and/or some intermediate outcomes and/or some time-dependent covariates.

To handle missing values in baseline covariates, we can expand the vector of baseline covariates, $X(0)$, to include the corresponding missing indicators, as proposed in Subsection 8.5.2.

To handle missing data in intermediate outcomes and/or time-dependent covariates, we can apply the multiple imputation (MI) procedure described in Subsection 5.1.1. In practice, we don't have to apply the MI procedure to impute all the missing values from $t = 1$ to $t = T$; instead, we only need to apply the MI procedure to impute the missing values to make the missing pattern become monotone and then apply the LTMLE approach to construct the main estimator for estimating the estimand.

Let $t_{\max}$ be the last time when the outcome variable (either primary outcome at T or intermediate outcome measured at $1 \leq t \leq T - 1$) is observed. Therefore, $1 \leq t_{\max} \leq T$. If there is any missing value between $t = 1$ and $t = t_{\max}$, then the missing pattern is non-monotone. We can describe the MI+LTMLE approach in three steps.

First, using a multiple imputation procedure to impute missing values in $X(s)$ and $Y(s)$ where s is between $t = 1$ and $t = t_{\max}$ such that $X(s)$ and $Y(s)$, $1 \leq s \leq t_{\max}$, are complete after imputation. Thus, after this step, the missing pattern becomes monotone.

Second, for the mth version of imputation, obtain a LTMLE estimate using the method discussed in Section 9.2.1, along with an estimate of its standard error, $m = 1, \ldots, M$.

Third, use Rubin's rule to combine the resulting estimates, providing a final estimate, along with an estimate of its standard error.

9.3 Implementation

We can implement LTMLE using R package "ltmle" (Lendle et al. 2017), by specifying either generalized linear model (GLM) or super learner for Q_t's, g_t's, and h_t's, assuming the missing is monotone. Some of the key arguments of R function "ltmle" are shown in what follows. Refer to Lendle et al. (2017) for the full syntax.

```
ltmle(
data,
Anodes,
Cnodes = NULL,
Lnodes = NULL,
```

```
Ynodes,
survivalOutcome = NULL,
Qform = NULL,
gform = NULL,
SL.library,
abar,
rule = NULL,
...
)
```

Here are some explanations of the above key arguments:

- `data`: data should be a data frame where the order of the columns corresponds to the time-ordering of the model, and the columns before (to the left of) the first of `Cnodes` or `Anodes` are treated as baseline covariates;
- `Anodes`: column names of treatment nodes that are corresponding to $A(0), \ldots, A(T-1)$;
- `Cnodes`: column names of censoring nodes that are corresponding to $1 - \Delta(1), \ldots, 1 - \Delta(T)$;
- `Lnodes`: column names of time-dependent covariate nodes corresponding to $X(1), \ldots, X(T-1)$;
- `Ynodes`: column names of outcome nodes corresponding to $Y(1), \ldots, Y(T)$ (If `survivalOutcome` is `FALSE`, `Ynodes` are treated as binary if all values are 0 or 1, and are treated as continuous otherwise);
- `survivalOutcome`: If `TRUE`, then `Ynodes` are indicators of an event (if $Y(t) = 1$ then $Y(t') = 1$ for ant $t' > t$);
- `Qform`: regression formula for outcome regression functions Q's;
- `gform`: regression formula for propensity score functions g's and non-missing probability functions h's;
- `SL.library`: libraries for fitting super learner to Q's, g's, and h's;
- `abar`: the counterfactual value of `Anodes` corresponding to $\overline{a}$;
- `rule`: specification of a dynamic treatment regime (`abar` and `rule` cannot both be specified. If one of them if a list of length 2, estimate of ATE can be computed).

We conclude this section with a simple example where there are two follow-up visits ($T = 2$) to demonstrate the use of R function "ltmle" to estimate $\theta_{\text{ATE}} = \nu(\overline{1}) - \nu(\overline{0})$. Assume that there are three baseline covariates at $t = 0$ (`L0.a`, `L0.b`, `L0.c`), two time-dependent covariates at $t = 1$ (`L1.a`, `L1.b`), and one outcome variable at $t = 1$ and $t = 2$ (`Y1`, `Y2`). Define the censoring variables at $t = 1$ and 2 (`C1`, `C2`), which are factor variables with two levels, "uncensored" or "censored." Table 9.1 displays the structure of the dataset to be defined in R.

In what follows is a sample of R codes presented in Lendle et al. (2017) used to implement the LTMLE estimator in the above example, providing the point estimate and 95% confidence interval of θ_{ATE}.

TABLE 9.1
Data structure in the example

Argument	Columns names
Baseline covariates	`c("L0.a", "L0.b", "L0.c")`
`Lnodes`	`c("L1.a", "L1.b")`
`Anodes`	`c("A0", "A1")`
`Cnodes`	`c("C1", "C2")`
`Ynodes`	`c("Y1", "Y2")`

```
data <- data.frame(L0.a, L0.b, L0.c, A0, C0, L1.a, L1.b, Y1, A1,
   C1, Y2)
Lnodes <- c("L1.a", "L1.b")
Anodes <- c("A0", "A1")
Cnodes <- c("C0", "C1")
Ynodes <- c("Y1", "Y2")
ltmle(data = data, Anodes = Anodes, Cnodes = Cnodes, Lnodes =
    Lnodes, Ynodes = Ynodes,  survivalOutcome = FALSE,
    abar = list(treament = c(1, 1), control = c(0,0)))
```

9.4 Exercises

Ex 9.1

Motivated by the proof in Subsection 9.1.3, show how to apply the shortcut of finding efficient influence function to obtain $\phi_{\text{EIF-H-}\overline{a}}(O)$ in (9.18).

Ex 9.2

1. Download the "ltmle" pdf from the following link:
   https://cran.r-project.org/web/packages/ltmle/ltmle.pdf
2. Run the R codes that come with R function"summary.ltmle."
3. Report the results they produce.

10

Sensitivity Analysis

10.1 Introduction

Let's summarize what we have achieved so far. Our goal is to answer a research question. In Chapter 1, we discussed how to define causal estimand reflecting the research question using the notion of potential outcomes. The next step is to design a study aligned with the causal estimand—either a randomized controlled clinical trial (RCT) or a real-world evidence (RWE) study.

Since the potential outcomes only exist in the potential world, we need to translate the causal estimand to a statistical estimand that is estimable in the real world. From Chapter 2 to Chapter 6, for each of the commonly used study designs, we discussed the task of identification, specifying a set of identifiability assumptions under which we are able to translate the causal estimand into a statistical estimand, and then discussed the task of estimation briefly—proposing a statistical method to estimate the corresponding statistical estimand. Note that in these chapters, for each statistical estimand, we only mentioned a simple statistical estimator to demonstrate that the statistical estimand is estimable, without discussing its properties or whether or not there are other estimators with better performance.

From Chapter 7 to Chapter 9, we discussed the art of estimation, focusing on commonly used methods (IPW, AIPW, MLE, TMLE, and LTMLE). For each method, we discussed the underlying model assumptions, discussing their asymptotic properties, and comparing their performance.

After we construct an estimator based on the selected estimation method and obtain data from the study, we plug the data into the estimator—note that an estimator is a function of the data—and obtain an estimate, along with its standard error estimate, confidence interval estimate, and p-value.

So far so good. Let's piece all these together. As shown in Figure 10.1, there are causal assumptions (i.e., the identifiability assumptions) under which we can translate the causal estimand into a statistical estimand and statistical assumptions under which we can construct an estimator with good properties. Then, we wish the estimate is a good estimate of the causal estimand. But, our wish depends on two bets: (1) the estimate is a good estimate of the statistical estimand under those statistical assumptions; and (2) the statistical estimand is equal to the causal estimand under those causal assumptions.

DOI: 10.1201/9781003433378-10

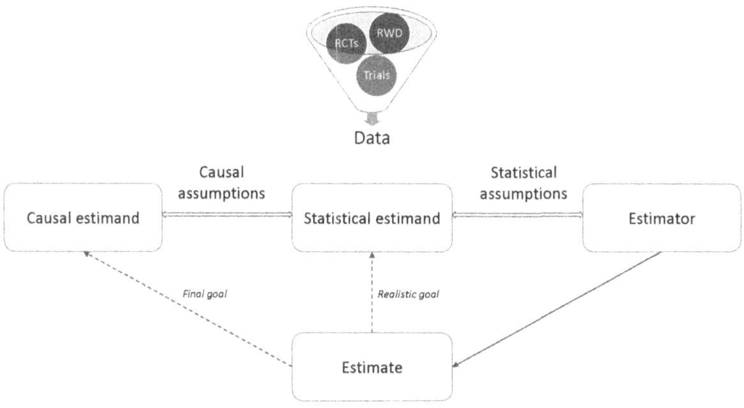

FIGURE 10.1
Underlying assumptions

However, assumptions are assumptions; they are not facts. Therefore, we need to evaluate what would happen if the assumptions were violated. This is the reason that *sensitivity analysis* is brought into play.

As seen from its title, ICH E9(R1) "Addendum on estimands and sensitivity analysis in clinical trials to the guideline on statistical principles for clinical trials" covered two main topics: estimand and sensitivity analysis. ICH E9(R1) provided a definition for sensitivity analysis.

> Definition of sensitivity analysis: "A series of analyses conducted with the intent to explore the robustness of inferences from the main estimator to deviations from its underlying modeling assumptions and limitations in the data."—ICH E9(R1)

Sensitivity analysis is different from supplementary analysis. ICH E9(R1) also provided a definition for supplementary analysis.

> Definition of supplementary analysis: "A general description for analyses that are conducted in addition to the main and sensitivity analysis with the intent to provide additional insights into the understanding of the treatment effect."—ICH E9(R1)

Chapters 1–6 of this book focus on estimand and Chapters 7–9 of this book focus on estimator. Now, Chapter 10 focuses on sensitivity analysis for causal assumptions.

Then what about sensitivity analysis for statistical assumptions? We would like to argue that it is not needed if we consider one estimation method that only assumes that the observations are independent and identically distributed (i.i.d.) but no other parametric model assumption. For example, consider a situation where there is one estimation method that assumes a parametric model and there is an alternative method that doesn't assume any parametric model. If such a dilemma were encountered, we would recommend the latter method that only makes the assumption that observations are i.i.d. but no other parametric model assumptions. Otherwise, if the former method that assumes a parametric model were used, then we would need to conduct sensitivity analysis to evaluate the robustness of the result if the underlying parametric model assumption is violated.

10.2 Sensitivity Analysis for Identifiability Assumptions

Consider a cohort study, producing a dataset consisting of $(X_i, A_i, Y_i), i = 1, \cdots, n$. Assume that the research objective is to estimate the following causal estimand,

$$\theta^* = \mathbb{E}\{Y^1\} - \mathbb{E}\{Y^0\}. \tag{10.1}$$

Under the identifiability assumptions (consistency, exchangeability, and positivity), we show that the above causal estimand is equal to the following statistical estimand,

$$\theta = \mathbb{E}\{\mathbb{E}(Y|X,1) - \mathbb{E}(Y|X,0)\}. \tag{10.2}$$

Then we can select one method with good properties, say the targeted minimum loss estimator (TMLE), which produces estimator $\widehat{\theta}_{\text{TMLE}}$. The reason that we prefer TMLE over other estimation methods is that TMLE is doubly robust, efficient, and stable. The only statistical assumption behind TMLE is that observations $(X_i, A_i, Y_i), i = 1, \cdots, n$, are i.i.d., without specifying any parametric model assumption for outcome regression function $Q(x, a)$ or the propensity score function $g(a|x)$. Therefore, we only need to conduct sensitivity analysis for those three identifiability assumptions.

10.2.1 The consistency assumption

The consistency assumption assumes that: if $A = 1$, then $Y = Y^{a=1}$; if $A = 0$, then $Y = Y^{a=0}$. If the consistency assumption is violated, then for some subjects with $A = 1$, $Y \neq Y^{a=1}$; while for some subjects $A = 0$, $Y \neq Y^{a=0}$. This violation may come from the discrepancy between the observed treatment A and the true treatment A^* received by some subjects.

The research on sensitivity analysis for the violation of the consistency assumption is lacking. A naive method is to swap a proportion of subjects between arm $A = 1$ and arm $A = 0$ (Fang and He 2023b). For example, we can randomly swap a range of percentages, say from 1% to k%, between two arms. For each percentage, we obtain an estimate—maybe conduct multiple random swaps and take an average—based on the data after swapping, producing $\widehat{\theta}^{(1)}, \ldots, \widehat{\theta}^{(k)}$, along with their statistical inferences.

As a side note, when considering an estimation method in the main analysis, the more sophisticated the better; while when considering a sensitivity analysis method, the more naive the better.

10.2.2 The exchangeability assumption

The exchangeability assumption is also known as the no-unmeasured-confounding (NUC) assumption. Therefore, we need to conduct sensitivity analysis to explore the robustness of the inference if there is unmeasured confounding. Although the literature on sensitivity analysis for consistency assumption is lacking, literature on sensitivity analysis for unmeasured confounding is rich. For example, Schneeweiss (2006) proposed several sensitivity analysis methods, Zhang et al. (2018) conducted a literature review, and Zhang, Stamey, and Mathur (2020) proposed a framework for doing sensitivity analysis for unmeasured confounding.

In this subsection, we briefly describe E-value (VanderWeele and Ding 2017) since it is a sensitivity analysis method without assumption. If we make an assumption to do sensitivity analysis, then we should do another layer of sensitivity analysis for the newly made assumption.

> "Definition of E-value: The E-value is the minimum strength of association, on the risk ratio scale, that an unmeasured confounder would need to have with both the treatment and outcome, conditional on the measured covariates, to fully explain away a specific treatment–outcome association."—VanderWeele and Ding (2017)

Let $RR_{AY|X}$ be the estimated relative risk of treatment A on binary Y conditional on measured confounders X, along with its 95% confidence interval (LL, UL), where LL is the lower limit and UL is the upper limit. Without loss of generality, we assume $RR_{AY|X} > 1$ and $LL > 1$. A caveat is that here the estimand of interest is the conditional relative risk, which is different from the estimands we have discussed in this book so far (e.g., the average treatment effect). For simplicity, here we omit this caveat (i.e., the difference between conditional estimands and marginal estimands) and focus on the application of E-value. Assume that there is one unmeasured confounder denoted as U. Since U is unmeasured, we need to conceptually define two parameters—one is the association of U and Y and the other is the association of U and A—to understand the true relative risk $RR_{AY|(X,U)}$. As in VanderWeele and Ding (2017), let RR_{UY} denote the maximum risk ratio for the outcome comparing any two categories of U, within either treatment group, conditional on X. Let RR_{AU} denote the maximum risk ratio for any specific level of U comparing $A = 1$ and $A = 0$, with adjustment already made for X.

VanderWeele, Ding, and Mathur (2019) provided the following technical definitions for RR_{UY} and RR_{AU}, conditional on $X = x$,

$$RR_{UY} = \max\left\{\frac{\max_u \mathbb{P}(Y = 1|A = 1, x, u)}{\min_u \mathbb{P}(Y = 0|A = 1, x, u)}, \frac{\max_u \mathbb{P}(Y = 1|A = 0, x, u)}{\min_u \mathbb{P}(Y = 0|A = 0, x, u)}\right\},$$

$$RR_{AU} = \max_u\left\{\frac{\mathbb{P}(U = u|A = 1, x)}{\mathbb{P}(U = u|A = 0, x)}\right\}.$$

The reader may skip the above expressions by focusing on the main idea.

The maximum relative amount by which the unmeasured confounders U could reduce the estimated relative risk $RR_{AY|X}$ is given by the following bias factor (Ding and VanderWeele 2016):

$$B = \frac{RR_{UY} \times RR_{AU}}{RR_{UY} + RR_{AU} - 1}. \tag{10.3}$$

To obtain the maximum amount that U could alter the estimated relative risk $RR_{AY|X}$, we can simply divide $RR_{AY|X}$ by B and divide the lower limit LL by B. Based on the definition of E-value (VanderWeele and Ding 2017), E-value is the minimum strength of association, on the relative risk scale, that U needs to have both the treatment and outcome, conditional on X, to explain away an estimated treatment effect. Ding and VanderWeele (2016) derived an explicit formula for E-value:

$$\text{E-value} = RR + \sqrt{RR(RR - 1)}, \tag{10.4}$$

where RR can be either the point estimate $RR_{AY|X}$ or the lower level LL.

Proof of E-value formula (10.4):

First, we show that $h(x, y) = xy/(x + y - 1)$ is an increasing function of x and y over domain $(1, \infty) \times (1, \infty)$. In fact, the derivative of $h(x, y)$ over x or y is positive over the domain. For example, the derivative of $h(x, y)$ over x is $y(y-1)/(x+y-1)^2$, which is positive over the domain. Among the solutions of RR_{UY} and RR_{AU} for inequality $RR/B \leq 1$, the minimum value of the RR_{UY} solutions is equal to the minimum value of the RR_{AU} solutions.

Then we examine inequality $RR/h(x, x) \leq 1$, which is equivalent to inequality $x^2 - 2RRx + RR \geq 0$ over domain $(1, \infty)$. Note that the solution to quadratic inequality $x^2 - 2RRx + RR \geq 0$ over domain $(1, \infty)$ is $x \geq RR + \sqrt{RR(RR - 1)}$. This proves the formula in (10.4). □

Table 10.1 shows some toy examples. For example, if the lower level of 95% confidence interval estimate is $LL = 1.33$, then E-value associated with this LL is 2.00. That means, to explain away such observed relative risk expressed as $LL = 1.33$, the magnitude of RR_{UY} and RR_{AU} (if $RR_{UY} = RR_{AU}$) should be at least as large as $1.33 + \sqrt{1.33(1.33 - 1)} = 2$.

TABLE 10.1
E-values for some toy examples

RR	$RR + \sqrt{RR(RR-1)}$	E-value
1.25	$1.25 + \sqrt{1.25(1.25-1)}$	1.81
1.33	$1.33 + \sqrt{1.33(1.33-1)}$	2.00
1.50	$1.50 + \sqrt{1.50(1.50-1)}$	2.37
1.75	$1.75 + \sqrt{1.75(1.75-1)}$	2.90
2.00	$2.00 + \sqrt{2.00(2.00-1)}$	3.41

E-value is defined on the relative-risk scale, but the estimated treatment effect doesn't have to be in terms of relative risk. Note that in the previous chapters, we often define the treatment effect in terms of risk difference for binary outcome and in terms of mean difference for continuous outcome. Linden, Mathur, and VanderWeele (2020) introduced R package "evalue," which can be used to calculate E-values associated with estimates in terms of relative risk, odds ratio, hazard ratio, risk difference, and mean difference.

10.2.3 The positivity assumption

The positivity assumption requires sufficient variability in the treatment variable within the strata of confounders. Positivity violations can arise for two reasons (van der Laan and Rose 2011). First, it may be impossible in the population level for subjects with certain covariate values to receive a given exposure of interest. If the positivity violation is due to this reason, we may redefine the estimand by restricting the population to some subpopulation. Second, violations or near violations of the positivity assumption may arise in finite samples due to chance, in particular in settings where the sample size is small or the set of covariates is large. In this subsection, we will describe a naive method for conducting sensitivity analysis to explore potential violation due to the second reason (Fang and He 2023b).

Let $\mathcal{S} = \{(X_{si}, A_{si}, Y_{si}), i = 1, \ldots, m\}$ consist of the data of subjects with propensity scores at the two extreme ends, say, with $\widehat{g}_{\mathrm{SL}}(1|X_i) > 1 - 0.05$ or $\widehat{g}_{\mathrm{SL}}(1|X_i) < 0.05$, where 0.05 is some pre-specified threshold to classify the potential positivity violation. Assume we consider the TMLE estimator,

$$\widehat{\theta}_{\mathrm{TMLE}} = \frac{1}{n}\sum_{i=1}^{n}\left\{\widehat{Q}^*(1, X_i) - \widehat{Q}^*(0, X_i)\right\}.$$

For each subject in $\mathcal{S}$, one of the two estimates in the summand of the above formula, $\widehat{Q}^*(1, X_i)$ and $\widehat{Q}^*(0, X_i)$, is not supported by data due to the positivity violation and therefore requires extrapolation, resulting in some degree of bias. To explore the magnitude of such bias, we may add some disturbance to the above extrapolation for subjects in $\mathcal{S}$. If the bigger value in Y means better treatment performance, then a positive estimate $\widehat{\theta}_{\text{TMLE}}$ indicates a favorable treatment effect. Hence, we subtract a positive disturbance δ from $\widehat{Q}^*(1, X_i) - \widehat{Q}^*(0, X_i)$ for each subject in $\mathcal{S}$ to manifest the unfavorable extrapolation bias, leading to an estimate after potential bias is removed,

$$\text{ß}\widehat{\theta}(\delta) = \widehat{\theta} - m\delta/n.$$

We may solve δ_0 such that $\widehat{\theta}(\delta_0) = 0$, which is similar to E-value or the tipping point method in the literature of missing data. The bigger δ_0 the more robust to the positivity violation. If δ_0 is unrealistically large, we may conclude the finding is trustful even under the positivity violation.

10.3 Sensitivity Analysis for the MAR Assumption

Sensitivity analysis for missing data handling when the MAR assumption is violated is rich; e.g., Mallinckrodt et al. (2020). There are two popular methods: the tipping-point method and the reference-based method.

In this section we consider longitudinal study in general. To handle missing data, we derive main results assuming missing at random (MAR) and then conduct sensitivity analysis for missing not at random (MNAR).

In the literature, several reference-based imputation methods, including "jump to reference" (J2R) and "copy reference" (CR), have been commonly used for conducting sensitivity analysis under MNAR. Here reference means the placebo arm (or the control arm in general) in RCTs or the standard of care cohort (or the control cohort in general) in cohort studies. J2R assumes the mean effect profile of patients who discontinue the investigative treatment jumps to that of the patients in the reference group after discontinuation, while CR assumes the conditional mean effect profile given the status prior to the time of discontinuation copies that of the patients in the reference group after discontinuation. Fang and Jin (2022) proposed a class of reference-based imputation methods, which includes J2R and CR as two extreme ends.

Multiple imputation (MI) is a popular method for handling missing data (Rubin 2004). The three basic steps of MI are: (1) imputation; (2) analyzing each completed dataset; and (3) combining results by Rubin's rule. The key step is the first step by which the missing data are imputed using a model-based imputation. In the following, we describe a class of reference-based imputation models (Fang and Jin 2022).

10.3.1 A class of reference-based imputation models

Recall that "missing" data are data that are considered as irrelevant due to the handling of intercurrent events (ICEs) regardless of being collected or not.

Let L denote the last visit before missing or "missing" occurs,

$$L = \arg\min\{t : \Delta_t = 0, \Delta_{t+1} = 1\}. \tag{10.5}$$

Imputation model $\mathcal{M}(0)$

Assume that missing or "missing" data are the consequence of an ICE, which is to be handled by the hypothetical strategy. We envisage a hypothetical scenario in which the investigative treatment gives up its effect completely after the ICE occurrence. Under this scenario, the post-ICE profile of the outcome variable would "jump" to the reference, so this imputation model is equivalent to the J2R imputation model.

Imputation models $\mathcal{M}_0(s), s = 0, \ldots, T-1$

Note that the imputation model $\mathcal{M}(0)$ or the J2R imputation model is conservative. Hence, we can make it less conservative by envisaging a hypothetical scenario in which the investigative treatment has partially carried-over (CO) effect from baseline $t = 0$ up to follow-up visit s or up to the last visit L, whichever is reached first. The imputation model under this scenario is referred to as $\mathcal{M}_0(s)$, $s = 0, \ldots, T-1$.

On the one extreme end, if $s = 0$, $\mathcal{M}_0(s)$ is equivalent to $\mathcal{M}(0)$. Therefore, this class of imputation models includes the J2R imputation model as a special case. On the other extreme end, the hypothetical scenario behind $\mathcal{M}_0(s)$ when $s = T-1$ is that the investigative treatment has CO effect from baseline $t = 0$ up to the last visit L. Hence, imputation model $\mathcal{M}_0(T-1)$ is equivalent to the CR imputation model. Therefore, this class of imputation models includes the CR imputation model as a special case.

We summarize these imputation models in Table 10.2.

Imputation model $\mathcal{M}(T)$

We can continue the above reasoning to envisage the least conservative hypothetical scenario, that is, the one in which the investigative treatment has full CO effect from baseline $t = 0$ up to the final follow-up visit $t = T$. We see that this imputation model is equivalent to the imputation model under

TABLE 10.2
A class of reference-based imputation models

Notation	Equivalence	Description/Assumption
$\mathcal{M}_0(0)$	J2R model	No CO effect at all; $0 \wedge L = 0$
...	...	...
$\mathcal{M}_0(s)$	$0 < s < T-1$	Partial CO effect up to $s \wedge L$
...	...	...
$\mathcal{M}_0(T-1)$	CR model	Partial CO effect up to $(T-1) \wedge L = L$

the MAR assumption. Therefore, if this imputation model is used to obtain the main estimator under the MAR assumption, we may use the above class of reference-based imputation models, $\mathcal{M}_0(s), s = 0, \ldots, T-1$, to conduct sensitivity analysis under the MNAR assumption.

10.3.2 Sequential modeling

Motivated by the sequential modeling for CR and J2R (O'Kelly and Ratitch 2014; Ratitch and O'Kelly 2011), Fang and Jin (2022) proposed the sequential modeling for implementing imputation models $\mathcal{M}_0(s)$, $s = 0, \ldots, T$. The implementation of $\mathcal{M}_0(s)$ consists of the following three steps:

Step 1: Create a tentative dataset from the original dataset, by tentatively deleting all the observed outcomes after visit s of subjects with missing data at visit T from both the treatment arm and the reference arm. This tentative dataset has more "missing" data (original missing data plus tentative "missing" data). Name this tentative dataset as IMPUTE0.

Step 2: Impute all the "missing" data (original missing data and tentative "missing" data) sequentially using the data of all the subjects in the reference arm and the data of those subjects with "missing" data in the treatment arm, producing a tentative completed dataset. Name this tentative completed dataset as IMPUTE1.

Step 3: Replace the imputed data in this completed tentative dataset by the originally observed data at the corresponding visits where tentative "missing" data are located, producing a final completed dataset. Name this final completed dataset as IMPUTE2.

After dataset IMPUTE2 is obtained, apply TMLE to obtain a point estimate. Repeat the process by M times. Combine the results by Rubin's rule. This completes the description of the MI procedure.

See Fang and Jin (2022) for more detail, along with some illustrative cartoons, key programming codes, and some numerical examples. These imputation methods are applicable to continuous outcomes, binary outcomes, and time-to-event outcomes.

10.4 Appendix

In Subsection 10.2.2, we use bounding factor B in (10.3) to calculate E-value to conduct sensitivity analysis for the NUC assumption. In this appendix, we will use bounding factor B to calculate sample size in the planning of observational studies (Fang et al. 2021).

In an RCT where treatment arms are assigned randomly, an unadjusted estimator, say $\widehat{RR}_{AY}$, is an unbiased estimator of the average treatment effect (ATE) of A on Y, measured by the relative risk $\theta = \mathbb{P}(Y^{a=1})/\mathbb{P}(Y^{a=0})$. Without loss of generality, assume that we wish $RR_{AY} > 1$, which means that the treatment $A = 1$ increases the success rate. Otherwise, we change the coding system for A by letting $A' = 1 - A$.

In an observational study, in the presence of confounder X, $\widehat{RR}_{AY}$ is a biased estimator of θ. The bias caused by confounder X can be bounded by a bounding factor B, similar to (10.3),

$$B = \frac{RR_{XY} \times RR_{AX}}{RR_{XY} + RR_{AX} - 1}.$$

In the analysis stage of an observational study, RR_{AX} and RR_{XY} can be estimated. Thus, to adjust for confounder X, we can shift the point estimate from unadjusted estimate $\widehat{RR}_{AY}$ to $\widehat{RR}_{AY}/B$ and its 95% confidence interval estimate, say (LL, UL), to $(LL/B, UL/B)$. If $LL/B > 1$, it means the adjusted 95% confidence interval doesn't cover 1 and we can claim that ATE is still statistically significant after adjusting for X.

However, in the planning stage of an observational study, RR_{AX} and RR_{XY} are unknown and there are no data yet to estimate them. Then, how do we do sample size calculation, taking into account the impact of confounder X? For this aim, we will have some guesses of RR_{AX} and RR_{XY}, and for each guess of RR_{AX} and RR_{XY}, we do the sample size calculation. We call this sensitivity analysis in the sample size calculation. We now describe such sensitivity analysis in detail in the following.

Assume that the investigator expects that the proportion of $Y = 1$ in cohort of $A = 0$ is p_0, and the relative risk of $Y = 1$ in cohort $A = 1$ versus cohort $A = 0$ is RR_{AY}. If we ignore the confounding, we can easily calculate a sample size that is needed to detect the expected relative risk RR_{AY} using a two-sided Z-test based upon $\log(RR_{AY})$ with 80% power under 5% significance level, using the formula (4.60) in Chapter 4. For example, assume the ratio of the sizes of two cohorts is $r = 1$, $p_0 = 20\%$, and $RR_{AY} = 2$, then $N = n_1 + n_0 = 80 + 80 = 160$.

Next, assume there is a binary confounder X and the investigator expects that $RR_{AX} = 2$ and $RR_{XY} = 1.5$. The unadjusted estimator is biased due to the presence of confounder X and the bias is bounded by $B = (2 \times 1.5)/(2 + 1.5 - 1) = 1.2$. Then we can easily calculate a sample size that is needed to detect the expected relative risk $RR^*_{AY} = RR_{AY}/B = 2/1.2 = 1.67$ using a two-sided Z-test stratified over levels of X with 80% power under 5% significance level, using the same formula (4.60), and the needed sample size becomes $N^* = 170 + 170 = 340$. Repeating the process, we can obtain the needed sample sizes for a range of values of RR_{AX} and RR_{XY}.

10.5 Exercises

Download the pdf file of R package "EValue" from the following link:
https://cran.r-project.org/web/packages/EValue/EValue.pdf

Ex 10.1

For binary outcome, the estimand can be defined in terms of relative risk (RR), odds ratio (OR), or risk difference (RD). Run the R codes that come with the R function "evalues" in the document, report the R outputs, and interpret the corresponding E-values that are produced in the R outputs.

Ex 10.2

If the estimand is defined in terms of mean difference, consider R function "evalues.MD" to calculate the E-value. Run the R codes that come with the R function "evalues.MD" in the document and report the R output.

11

A Roadmap for Causal Inference

11.1 Introduction

Our journey starts with a research question, and we wish to arrive at the final destination successfully, generating evidence for decision-making on whether the investigative medical product is safe and beneficial. Thus, we need a roadmap for causal inference.

In Chapters 2–10, we discussed the following stops on the journey.

- Study type:
  - Randomized controlled clinical trial (RCT)
  - Real-world evidence (RWE) study
- Number of follow-ups:
  - Single follow-up
  - Multiple follow-ups
- Outcome type:
  - Continuous outcome
  - Binary outcome
  - Time-to-event outcome
- Intercurrent event (ICE) handling strategy:
  - Treatment policy strategy
  - Hypothetical strategy
  - Composite variable strategy
  - While on treatment strategy
  - Principal stratum strategy
- Missing data:
  - Missing data that are consequences of ICEs
  - Missing data that are not consequences of ICEs

DOI: 10.1201/9781003433378-11

- Type of estimand:
  - Intent-to-treatment (ITT) effect
  - Per-protocol (PP) effect
  - Average treatment effect (ATE)
  - Average treatment effect among the treated (ATT)
- Estimation method:
  - Inverse probability weighted (IPW)
  - Augmented IPW (AIPW)
  - Maximum likelihood estimator/minimum loss estimator (MLE)
  - Targeted MLE (TMLE)
- Sensitivity analysis for the following assumption:
  - Consistency assumption
  - Exchangeability assumption
  - Positivity assumption
  - Missing at random (MAR) assumption

Bearing in mind these stops discussed in the previous chapters, for any given research question, we are able to plan a trip to arrive at the final destination with the help of the roadmap for causal inference to be discussed.

11.2 Roadmap

Figure 11.1 displays a roadmap for causal inference. There are 9 steps in the roadmap, which can be categorized into four stages: (1) the study protocol stage, (2) the data collection stage, (3) the statistical analysis plan (SAP) stage, and (4) the clinical study report (CSR) stage.

11.2.1 Study protocol

Step 1: Research question

In Chapter 1, we discussed the central questions for drug development and licensing. We also discussed how to specify a sound research question using the PROTECT criteria.

The following two steps (i.e., Step 2 and Step 3) are not in chronological order; they should be aligned with each other.

Step 2: Study design

In the previous chapters, we discussed RCTs, which include but are not limited to complete RCTs and stratified RCTs, and RWE studies, which include but are not limited to pragmatic RCTs, externally controlled trials (ECTs), and observational studies.

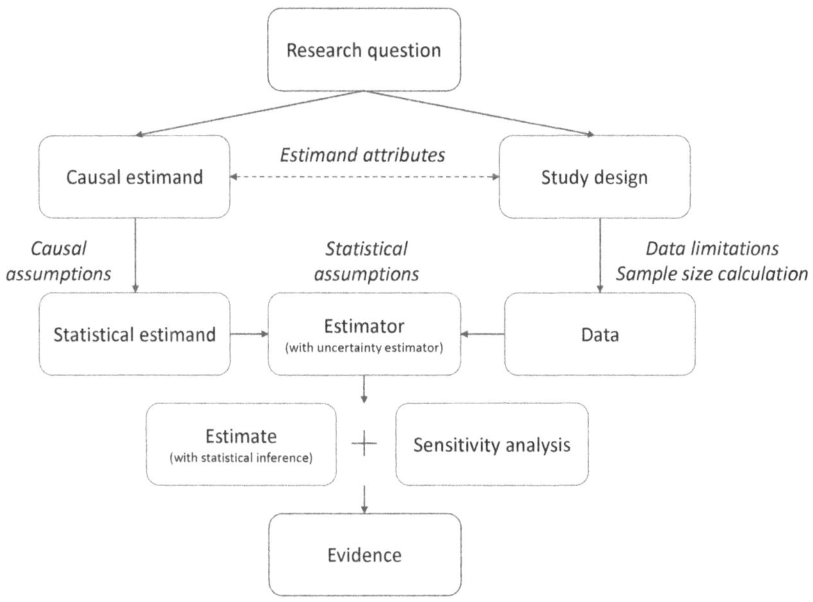

FIGURE 11.1
A roadmap for causal inference

After we select one study design, in the study planning stage, we need to assess the study design and discuss the potential data limitations. For example, if we select an RCT design, we need to assess whether there are any anticipated missing data, whether there are any anticipated ICEs, and which missing data are consequences of which ICEs. As another example, if we select an observational design based on preexisting real-world databases, we need to assess data availability, data relevance, and data quality.

A key component of study design is sample size calculation and power analysis, which should be aligned with the causal estimand. In Section 4.4, we demonstrated that sample size calculation should be aligned with how missing data and ICEs will be handled, which is one of the five attributes of estimand. In Section 10.4, we demonstrated that sample size calculation should also be aligned with the causal estimand in the presence of potential confounding bias.

Step 3: Causal estimand

In Chapter 1, we discussed how to define a causal estimand to reflect the research question, using the notion of potential outcomes. In Chapters 2–3, we discussed how to follow ICH E9(R1) to think through the five attributes of the estimand: population, treatment, outcome variable, how to deal with anticipated missing data and ICEs, and population-level summary.

After defining the causal estimand, we move forward to the task of identification. For this task, we should specify identifiability assumptions that are aligned with the study design. In many scenarios, we make the following assumptions: (1) the consistency assumption; (2) the assumptions embedded in the directed acyclic graph (DAG) that is aligned with the study design; (3) the positivity assumption; and (4) the underlying assumptions for handling missing data and ICEs.

11.2.2 Data collection

Step 4: Data

Data are either generated from the study or are preexisting. Specifically, there are prospective study designs, such as RCTs and prospective observational studies, which generate data. There are also retrospective study designs, which are designed based on some preexisting real-world databases. In addition, there are hybrid studies, in which there are prospective components and retrospective components.

Although this step is out of the scope of statistical work, statisticians are often consulted in the process of data collection, to ensure that the data collection process is guided by the study protocol. Statisticians need to be informed and consulted when evaluating whether there is any protocol deviation. If there is any protocol deviation that cannot be fixed, the study protocol needs to be amended. Consequently, we need to revisit the previously specified DAG and identifiability assumptions to see if they should be revised.

11.2.3 Statistical analysis plan

Step 5: Statistical estimand

This step is to continue Step 3. Under the identifiability assumptions, we need to complete the task of identification. In Chapter 6, we discussed two basic strategies for translating the causal estimand into a statistical estimand: the standardization strategy and the weighting strategy.

Step 6: Estimator

This step is to propose an estimation method to estimate the statistical estimand developed in the preceding step. In Chapters 7–9, we discussed several estimation methods, assessed the underlying statistical assumptions behind these methods, and compared their asymptotic properties. From these methods, we select the most appropriate method, which produces a point estimator, along with an estimator of its standard error.

In addition to the main analysis plan, in this step, we need to propose a plan of sensitivity analysis to evaluate the robustness of the main results if the underlying assumptions are violated.

Estimand, *estimator*, and *sensitivity analysis* are the three key components of any well-developed SAP. These three terms form the title of the monograph (Mallinckrodt et al. 2020).

"Defining estimands is making sure we do the right things; constructing estimators is making sure we do things right; conducting sensitivity analysis is making sure we do the right things right."—Fang (2020)

11.2.4 Clinical study report

The data collection stage and the SAP development stage are not in chronological order; they are two parallel stages. After the SAP has been approved and after the database lock, we execute the SAP using the data.

Step 7: Estimate

An estimator is a function of data. Thus, the main estimator proposed in Step 6 is a function of data collected in Step 5, so is the estimator of its standard error. After we obtain the data, we plug the data into the main estimator and the estimator of its standard error, we obtain a point estimate along with an estimate of its standard error. Based on these estimates, we can conduct statistical inference, obtaining a confidence interval estimate and/or p-value.

Step 8: Sensitivity analysis

After we obtain the results from the above main analysis, following the sensitivity analysis plan proposed in the SAP, we conduct sensitivity analysis.

Step 9: Evidence

The last step is to interpret the results of the main analysis and the corresponding sensitivity analysis. The findings from such interpretation are summarized as evidence.

11.3 A Plasmode Case Study

We consider a case study based on a plasmode dataset that was simulated based on a subset of the NHEFS study (National Health and Nutrition Examination Survey Data I Epidemiologic Follow-up Study). A detailed description of the NHEFS study can be found at www.cdc.gov/nchs/nhanes/nhefs. The subset of the NHEFS data (including subject ID, outcome variable, and 9 baseline variables) was prepared by Hernán and Robins for their book (Hernán and Robins 2020). The dataset can be downloaded from their book website, https://www.hsph.harvard.edu/miguel-hernan/causal-inference-book. Hernán and Robins (2020) used the dataset to showcase the methods and the analyses discussed in their book and stated the following statement: "encourage readers to improve upon and refine our analyses."

11.3.1 Research question

Hypothetically, assume we want to investigate the long-term effectiveness of a made-up drug (MUD) compared with the standard of care (SOC) among a population of subjects who satisfy a set of inclusion and exclusion criteria. The treatment duration of both MUD and SOC is one year. The outcome variable is the body weight change measured at 11 years from the treatment initiation (i.e., 10 years after the end of the treatment).

The research question is: What are the treatment effects of treatment MUD vs. treatment SOC on the body weights measured at 11 years after the treatment initiation among all the subjects in the population?

Note that the above research question has all five components of the PROTECT criteria: population, response/outcome (i.e., body weight), treatment/exposure (i.e., MUD vs. SOC), counterfactual thinking (i.e., potential outcomes had all the subjects in the population been treated by MUD; potential outcomes had all the subjects in the population been treated by SOC), and time (i.e., 11 years after treatment initiation).

11.3.2 Study design

To investigate the long-term effectiveness, we decide to design a retrospective cohort study based on some preexisting database, after assessing the availability, relevance, and quality of the database.

The design is shown schematically in Figure 11.2. It is a cohort study consisting of two cohorts, MUD and SOC. The baseline time-point is the time when either MUD or SOC is initiated. Although the absolute baseline time-points are different among subjects within a time-period (e.g., from January 1, 1971 to December 31, 1971), they are all marked as time $t = 0$ in the schema. Note the schema doesn't show the present time—if the present time is before the baseline time, the study is prospective, if the present time is after final measurement time T, the study is retrospective, and if the present time is between 0 and T, the study is a hybrid study that is both retrospective and prospective. In this case study, the study is a retrospective study.

As indicated in the schematic design, there may be ICEs that occur after treatment initiation and before the outcome variable is measured. In this retrospective study, assume that we anticipate that some subjects may have treatment dropout (e.g., early termination of the treatment) and some subjects may have analysis dropout (i.e., missing data on the outcome variable).

FIGURE 11.2
A schematic illustration of the retrospective cohort study

11.3.3 Causal estimand

Let Y be the body weight change from baseline to $T = 11$ years. Let A be the treatment variable, with $A = 1$ standing for MUD and $A = 0$ for SOC. Let X be the vector of pre-treatment covariates. There are two anticipated ICEs, treatment dropout and analysis dropout. We propose to apply the treatment policy strategy to handle treatment dropout, considering treatment dropout as a part of the treatment, and apply the hypothetical strategy to handle analysis dropout, envisaging a hypothetical scenario in which analysis dropout would not occur. Let Δ be the analysis dropout status at T, with $\Delta = 1$ standing for the occurrence of dropout and $\Delta = 0$ indicating no analysis dropout. Let $Y^{a=j,\delta=0}$ be the potential outcome of Y if the subject had been treated by $A = j$ and there had been no analysis dropout, $j = 1, 0$.

Let n be the sample size. Assume that $(X_i, A_i, \Delta_i, (1 - \Delta_i)Y_i)$, $i = 1, \ldots, n$, are i.i.d. observations with $(X, A, \Delta, (1 - \Delta)Y)$ in the real world and that $(Y_i^{a=1,\delta=0}, Y_i^{a=0,\delta=0})$, $i = 1, \ldots, n$, are i.i.d. potential outcomes with $(Y^{a=1,\delta=0}, Y^{a=0,\delta=0})$ in the potential world.

Since the outcome variable is continuous, we consider the population mean as the population-level summary. After clarifying all five attributes (population, outcome, treatment, how to handle ICEs, and summary), we define the following causal estimand that reflects the research question,

$$\theta^* = \mathbb{E}(Y^{a=1,\delta=0}) - \mathbb{E}(Y^{a=0,\delta=0}). \tag{11.1}$$

We make the following assumptions for the task of identification:

1. The consistency assumption:

$$Y = Y^{a=j,\delta=0} \text{ if } A = j \text{ and } \Delta = 0, \text{ for } j = 1, 0;$$

2. The exchangeability assumption:
$$(Y^{a=0,\delta=0}, Y^{a=1,\delta=0}) \perp\!\!\!\perp A | X;$$

3. The missing at random (MAR) assumption:
$$(Y^{a=0,\delta=0}, Y^{a=1,\delta=0}) \perp\!\!\!\perp \Delta | (X, A);$$

4. The positivity assumption:
$$\mathbb{P}(A = j, \Delta = 0 | X = x) > 0, \text{ for } j = 1, 0; x \in \text{supp}(X).$$

11.3.4 Data

As mentioned earlier, we created a plasmode dataset based on the NHEFS dataset used by Hernán and Robins (2020). First, we obtain the vector of 9 pre-treatment variables measured in 1971 from the NHEFS dataset, which are (1) sex (0: male, 1: female), (2) race (0: white, 1: other), (3) age, (4) education (1: 8th grade or less, 2: high school dropout, 3: high school, 4: college dropout, 5: college or more), (5) cigarettes per day, (6) years of smoking, (7) exercise (0: much exercise, 1: moderate exercise, 2: little or no exercise), (8) active (0: very active, 1: moderately active, 2: inactive), and (9) weight in kg in 1971. We import the dataset in R and name the data.frame as `data0`.

Second, we rename variable `qsmk` in the NHEFS dataset as A, with `qsmk=1` replaced by $A = 1$ and `qsmk=0` replaced by $A = 0$, respectively. There are $n_1 = 403$ subjects in the cohort of $A = 1$ (a.k.a., the treated cohort, the MUD cohort) and $n_0 = 1163$ subjects in the cohort of $A = 0$ (a.k.a., the control cohort, the SOC cohort).

Third, we simulate an missing indicator variable, Δ, with $\Delta = 1$ standing for analysis dropout, using the following logistic model,

$$\text{logit}\{\mathbb{P}(\Delta = 1)\} = 0.5 \times A - \texttt{age}/50 - 1.5 \times (\texttt{wt71}/100)^2. \tag{11.2}$$

Fourth, we rename variable `wt82_71` in the NHEFS dataset as Y, which is the outcome variable defined as the body weight in kg in 1982 minus that in 1971, and set the value of Y as missing if $\Delta = 1$.

The R codes to implement the above steps are:

```
set.seed(1)
# create variable A
data0$A <- data$qsmk
# simulate missing status D
D <- rbinom(nrow(data0), size=1, prob=plogis(0.5*data0$A
                - 2*(data0$age/100)-1.5*(data0$wt71/100)^2))
# create outcome variabe Y with missing data
Y <- ifelse(D, NA, data0$wt82_71)
```

We can summarize those variables in the plasmode dataset, using mean for continuous variables and proportion for binary variables, overall and by cohorts (i.e., by $A = 1$ and $A = 0$), respectively. We omit the results here and relegate them to one of the exercises.

11.3.5 Statistical estimand

We are able to translate the causal estimand (11.1) into a statistical estimand under those identifiability assumption. In fact,

$$\begin{aligned}\theta^* &= \mathbb{E}(Y^{a=1,\delta=0}) - \mathbb{E}(Y^{a=0,\delta=0}) \\ &\overset{(a)}{=} \mathbb{E}[\mathbb{E}(Y^{a=1,\delta=0}|X)] - \mathbb{E}[\mathbb{E}(Y^{a=0,\delta=0}|X)] \\ &\overset{(b)}{=} \mathbb{E}[\mathbb{E}(Y^{a=1,\delta=0}|A=1,X)] - \mathbb{E}[\mathbb{E}(Y^{a=0,\delta=0}|A=0,X)] \\ &\overset{(c)}{=} \mathbb{E}[\mathbb{E}(Y^{a=1,\delta=0}|A=1,\Delta=0,X)] - \mathbb{E}[\mathbb{E}(Y^{a=0,\delta=0}|A=0,\Delta=0,X)] \\ &\overset{(d)}{=} \mathbb{E}[\mathbb{E}(Y|A=1,\Delta=0,X)] - \mathbb{E}[\mathbb{E}(Y|A=0,\Delta=0,X)] \triangleq \theta,\end{aligned}$$

where (a) is using law of iterated expectations, (b) holds under the exchangeability and positivity assumptions, (c) holds under the MAR and positivity assumptions, and (d) holds under the consistency assumption.

11.3.6 Estimator

In order to estimate the following statistical estimand,

$$\theta = \mathbb{E}[\mathbb{E}(Y|A=1,\Delta=0,X)] - \mathbb{E}[\mathbb{E}(Y|A=0,\Delta=0,X)], \tag{11.3}$$

we propose to consider the TMLE method, which is doubly robust and asymptotically efficient.

11.3.7 Estimate

To implement TMLE, we need to specify a super learner (SL) library for the outcome regression function, $Q(A,X) = \mathbb{E}(Y|\Delta=0,A=1,X)$, a SL library for the propensity score function, $g(1|X) = \mathbb{P}(A=1|X)$, and a SL library for the non-missing probability function, $h(A,X) = \mathbb{P}(D=0|A,X)$.

In the following R codes, we consider a SL library, which consists of generalized linear model (`SL.glm`), classification and regression tree model (`SL.rpart`), GLM regularized by elastic net (`SL.glmnet`), and random forest (`SL.randomForest`), for $Q(A,X)$ and $g(1|X)$; while we consider a slightly different SL library, which consists of `SL.rpart`, `SL.glmnet`, and generalized additive model (`SL.gam`), for $h(A,X)$.

```
# delta=0 means missing and delta=1 means no missing
data0$delta <- 1-data0$D

# Q.SL.library is for outcome-regression function
# g.SL.library is for propensity-score function
# g.Delta.SL.library is for non-missing probability function
# X.names is vector of names of covariates in X
tmle_fit <- tmle(Y=data0$Y, A=data0$A, W=data0[, X.names], Delta=
    data0$delta, Q.SL.library = c("SL.glm", "SL.rpart", "SL.glmnet
```

```
    ", "SL.randomForest"), g.SL.library = c("SL.glm", "SL.rpart",
    "SL.glmnet", "SL.randomForest"), g.Delta.SL.library = c("SL.
    rpart", "SL.glmnet", "SL.gam"), family = "gaussian")

nhefs_tmle_fit.h$estimates$ATE$psi
nhefs_tmle_fit.h$estimates$ATE$CI
```

From the output produced by the above R codes, we see that the point estimate of θ is 3.28, with SE=0.50 and 95% CI=$(2.31, 4.28)$.

11.3.8 Sensitivity analysis

Recall that we make four assumptions (consistency, exchangeability, MAR, and positivity) for the task of identification.

Let's examine the consistency assumption first. In Step 4, assume the data have good quality, with accurate data on the treatment variable and the outcome variable. In addition, we apply the treatment policy strategy to handle treatment dropout. Therefore, it is fine to assume that the consistency assumption is reasonable and skip the sensitivity analysis for it.

Next, let's examine the positivity assumption. Figure 11.3 shows the propensity score densities associated with $g(1|X)$, where group 1 is the MUD cohort and group 0 is the SOC cohort. We see that the propensity score densities have a lot of overlapping between the two cohorts and are bounded by 0.04 from the below and 0.8 from the above. It indicates that there may be no serious positivity violation. We do similar checking for $h(A, X)$ and obtain similar findings. Therefore, it is fine to assume that the positivity assumption is reasonable and skip the sensitivity analysis for it.

Then, let's examine the exchangeability assumption. We apply E-value (VanderWeele and Ding 2017) to conduct sensitivity analysis to explore the robustness of the main results if there is unmeasured confounding. E-Value

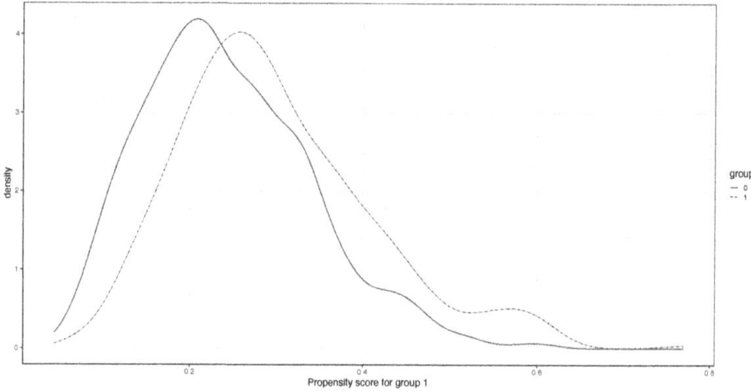

FIGURE 11.3
The propensity score densities (group 1 is MUD; group 0 is SOC)

was originally proposed for a binary outcome. For continuous outcome (Linden, Mathur, and VanderWeele 2020), we first calculate the point estimate of the standardized mean difference (i.e., Cohen's d), using the standard deviation of the outcome variable, which is equal to SD=8.03. Then the point estimate of Cohen's d is 3.28/7.88=0.4085 with SE=0.50/8.03=0.0622. We use R function "evalues.MD" of R package "EValue" to calculate the E-value, with the following R codes,

```
> evalues.MD(est=0.4085, se=0.0622)
```

It shows that the E-value corresponding to the point estimate is 2.26 and the E-value corresponding to the lower end of 95% CI is 1.92. Thus, in order to explain away the estimated treatment effect, unmeasured confounder(s) would need to approximately double the probability of a subject's being exposed to treatment MUD vs. SOC, and would need to approximately double the probability of being high vs. low on the outcome. This sensitivity analysis shows that the estimated treatment effect is robust.

Last, let's examine the MAR assumption. We consider reference-based imputation methods (O'Kelly and Ratitch 2014). Since there is only one follow-up time-point (i.e., $T = 11$ years), copy-reference (CR) and jump-to-reference (J2R) are equivalent. To implement them, we use R package "mice" to generate 100 imputations, with the following R codes,

```
# create a subset consisting of all data in cohort $A=0$ or
    oberved data in cohort $A=1$
data.ref0<-data0[(data0$A==0)|(data0$delta==0), ]
data.ref <-data.ref0
data.ref[, c("Y", "wt82", "A", "delta")] <- NULL
data.refim <- mice(data.ref, m=100, seed=500)
```

Based on each completed dataset, we obtain one TMLE estimate. Then we use Rubin's rule to combined 100 TMLE estimates, providing a pooled estimate, which is 3.35 with SE=0.49. We see that the estimate under CR or J2R is similar to the estimate under the MAR assumption. Hence the result is robust under the CR and J2R scenarios.

11.3.9 Evidence

The TMLE estimate of the statistical estimand is $\widehat{\theta}_{\text{TMLE}} = 3.28$, with SE=0.50 and 95% CI=$(2.31, 4.28)$. The results of sensitivity analysis show that the results of the main estimator are robust. Therefore, there is strong evidence to state that the estimated treatment effect is statistically significant, and the evidence is robust.

11.4 Exercises

Ex 11.1

Continue the plasmode case study in Section 11.3. Dichotomize continuous variable `wt82_71` into into a binary outcome, setting $Y_{\text{bin}} = 1$ if `wt82_71` > 5kg and $Y_{\text{bin}} = 0$ if `wt82_71` ≤ 5kg.

Ex 11.2

Continue the plasmode case study with binary outcome Y_{bin}. Summarize those pre-treatment variables and the newly defined binary outcome, using mean for continuous variables and proportion for binary variables, overall and by cohorts (i.e., by $A = 1$ and $A = 0$), respectively.

Ex 11.3

Continue the plasmode case study with binary outcome Y_{bin}. Report the TMLE estimates for the ATE estimand and the ATT estimand.

Ex 11.4

Continue the plasmode case study with binary outcome Y_{bin}. Report the E-values corresponding to the above TMLE estimates and interpret them.

12

Applications of the Roadmap

12.1 Introduction

In the preceding chapter, we proposed a detailed nine-step roadmap for causal inference, from specifying the research question to deriving evidence. In this chapter, we will present several applications of the roadmap.

Throughout this book, we focus on the central questions for drug development and licensing, "to establish the existence, and to estimate the magnitude, of treatment effects: how the outcome of treatment compares to what would have happened to the same subjects under alternative treatment," as stated in ICH E9(R1).

After specifying a sound research question, we need to select a study design. There are two major categories of study designs: traditional randomized controlled clinical trials (RCTs) and real-world evidence (RWE) studies. Within each category, there are many commonly used study designs from which we can select one. In the following two sections, we will discuss some applications of the roadmap to RCTs and to RWE studies, respectively.

12.2 Applications to RCTs

Consider a sample of subjects to be enrolled in a randomized controlled clinical trial (RCT). This sample is also known as the intent-to-treat (ITT) population, which is assumed to be sampled from a super-population that is conceptually constructed. Therefore, as in Chapters 2–5, we denote the sample size (i.e., the size of the ITT population) as N.

The main difference between an interventional study and a non-interventional study is that the propensity score function is known in the interventional study. For example, in an RCT, if the randomization ratio is r:1 between two arms, $Z = 1$ and $Z = 0$, then $\mathbb{P}(Z = 1) = r/(1 + r)$ for each subject. In a stratified RCT, if the stratification factor is S, with $S = 1, \ldots, K$, and the randomization ratio is r_k:1 within stratum $S = k$, then $\mathbb{P}(Z = 1|S = k) = r_k/(1 + r_k)$.

DOI: 10.1201/9781003433378-12

12.2.1 RCTs with a single follow-up

Step 1: Research question

Follow the PROTECT criteria to prepare the following items:

- Population: defined via inclusion/exclusion criteria;
- Response/Outcome variable: Y;
- Treatment/Exposure variable: treatment 1 or 0;
- Counterfactual thinking: randomization and blinding;
- Time: outcome measured at T.

Research question/objective: to establish the existence, and estimate the magnitude, of treatment effects of treatment 1 vs. treatment 0 on the outcome variable measured at time T from the time of treatment initiation among all the subjects in the well-defined ITT population.

Step 2: Study design

Consider a complete RCT or a stratified RCT with stratification factor S. In a typical double-blind RCT, treatment 0 is placebo.

Step 3: Causal estimand

Following Chapter 2, define a causal estimand in terms of potential outcomes, assuming there is no missing data or intercurrent events (ICEs). Following Chapters 3–4, revise the causal estimand to take into account the handling of the anticipated missing data and ICEs.

Example 1. In a stratified RCT, assume that there are two types of ICEs, treatment dropout and analysis dropout, and assume that the hypothetical strategy—envisaging a hypothetical scenario in which no dropout would occur—is applied to handle both treatment dropout and analysis dropout. Let Z be the treatment assignment and let Δ be the indicator of dropout (either treatment dropout or analysis dropout) occurence. Assume we are interested in the following causal estimand:

$$\theta_1^* = \mathbb{E}[Y^{z=1,\delta=0}] - \mathbb{E}[Y^{z=0,\delta=0}].$$

Remark: As discussed in Chapter 4, if a different strategy (or a different combination of strategies) is applied to handle ICEs, then the research question in Step 1 needs to be revised accordingly. For example, if the treatment policy strategy is applied to handle treatment dropout and the hypothetical strategy is applied to handle analysis dropout, then the "T/E" component of the research question in Step 1 needs to be revised, incorporating the treatment dropout as a part of the treatments being compared. If the composite variable strategy is applied to handle treatment dropout and analysis dropout, then the "R/O" component of the research question in Step 1 needs to be revised, redefining the outcome variable as a composite outcome variable. If the while on treatment strategy is applied to handle treatment dropout and the composite variable strategy is applied to handle analysis dropout, then the "T" component and "R/O" component of the research question in Step 1 need to be revised, redefining the outcome variable as a composite outcome variable measured at a new time-point. If the principal stratum strategy is applied to handle both treatment dropout and analysis dropout, then the "P" component of the research question in Step 1 needs to be revised, replacing the original ITT population by the principal stratum of interest.

Step 4: Data

Conduct sample size calculation that is aligned with the causal estimand, using one of the methods discussed in Chapter 4. Then enroll patients. And then collect data following the study protocol.

Example 1 (continued). Collect data

$$\mathcal{O} = \{(X_i, Z_i, \Delta_i, (1-\Delta_i)Y_i), i = 1, \ldots, N\},$$

where X includes stratification factor S, covariates that ensure the missing at random (MAR) assumption be satisfied, and covariates that are believed to be associated with the outcome variable.

Step 5: Statistical estimand

Under the identifiability assumptions, translate the causal estimand into a statistical estimand. This is the task of identification.

Example 1 (continued). Under the identifiability assumptions (consistency, MAR, positivity), the causal estimand can be translated into the following statistical estimand,

$$\theta_1 = \mathbb{E}[Y|X, Z=1, \Delta=0] - \mathbb{E}[Y|X, Z=0, \Delta=0].$$

Step 6: Estimator

To estimate the statistical estimand, one of the estimation methods discussed in Chapters 7–8 is proposed.

Example 1 (continued). TMLE is doubly robust in the sense that it is consistent if either the non-missing probability function or the outcome regression function is estimated consistently. It is also a regular and asymptotically linear (RAL) estimator with the efficient influence function.

Step 7: Estimate

Plugging the data into the proposed estimator, a point estimate is obtained, along with an estimate of its standard error, 95% confidence estimate for estimating the magnitude of the treatment effect, and p-value for testing the existence of the treatment effect.

Example 1 (continued). To implement TMLE, consider the following R function "tmle."

```
tmle_fit <- tmle(Y = Y, A = Z, W = W, Delta = 1 - Delta, Q.SL.library = c("SL.glm", "tmle.SL.dbarts2", "SL.glmnet"), g1W = g1S, g.Delta.SL.library = c("SL.glm", "tmle.SL.dbarts.k.5", "SL.gam"), family = "gaussian", ...)

tmle_fit$estimates$ATE$psi
tmle_fit$estimates$ATE$CI
```

In the above R function, `Y=Y` is to assign outcome variable Y on the right to argument `Y` on the left, `A=Z` is to assign treatment assignment Z to argument `Z`, `W=W` is to assign the expanded covariate vector, W (including covariates in X and the corresponding missing-covariate indicators), to argument `W`, `Delta=1-Delta` is to assign 1 minus missing indicator Δ in the data (0 for non-missing and 1 for missing) to argument `Delta` (1 for non-missing and 0 for missing), `Q.SL.library` is to specify a library of predictive models to fit super learner for the outcome regression function, `g1W=g1S` is the vector of conditional treatment assignment probabilities, $g(1|S) = \mathbb{P}(Z = 1|S)$ for stratified RCT or $g(1|S) = 1/2$ for 1:1 complete RCT, `g.Delta.SL.library` is to specify a library of predictive models to fit super learner for the non-missing probability function, and `family` specifies "gaussian" for continuous outcome and "binomial" for binary outcome.

Step 8: Sensitivity analysis

Conduct sensitivity analysis to explore the robustness of the results from the main estimator to deviations from its underlying assumptions.

Example 1 (continued). The identifiability assumptions are the consistency, MAR, and positivity assumptions. For RCTs, the consistency assumption and the positivity assumption are usually not worrisome. For the MAR assumption, a series of reference-based imputation methods discussed in Chapter 10 can be carried out to explore the robustness of the results under some missing not-at-random (MNAR) scenarios.

Step 9: Evidence

The last step is to interpret the results of the main analysis and the corresponding sensitivity analysis. The findings from such interpretation are summarized as evidence.

Example 1 (continued). If the results show that the treatment effect is statistically significant under MAR and statistically significant under those pre-specified MNAR scenarios, then the evidence is robust. However, if the results show that the treatment effect is not statistically significant under those pre-specified MNAR scenarios, then the evidence is not robust.

12.2.2 Longitudinal RCTs

Step 1: Research question

Follow the PROTECT criteria to prepare the following items:

- Population: defined via inclusion/exclusion criteria;
- Response/Outcome variable: Y;
- Treatment/Exposure variable: treatment 1 or 0;
- Counterfactual thinking: randomization and blinding;
- Time: outcome measured at $t = 1, \ldots, T$.

Research question/objective: to establish the existence, and estimate the magnitude, of treatment effects of treatment 1 (throughout T decision points) vs. treatment 0 (throughout T decision points) on the primary outcome measured at time T among subjects in the well-defined ITT population.

The pair of treatments under comparison can be replaced by any other two feasible treatment regimes to be investigated by a SMART design discussed in Chapter 5.

Step 2: Study design

Consider a complete RCT or a stratified RCT with stratification factor S, with follow-up visits indicated by $1, \ldots, T$. In a typical double-blind RCT, treatment 0 is a placebo.

Step 3: Causal estimand

Following Chapter 5, define a causal estimand in terms of potential outcomes, taking into account the handling of the anticipated missing data and ICEs.

Example 2. In a stratified RCT, assume that there are two types of ICEs, treatment dropout and analysis dropout, and assume that the hypothetical strategy—envisaging a hypothetical scenario in which no dropout would occur—is applied to handle both treatment dropout and analysis dropout. Let Z be the treatment assignment and let $\overline{\Delta} = (\Delta(1), \ldots, \Delta(T))$ be the vector of the indicators of dropout occurrence at T measurement time-points. Assume we are interested in the following causal estimand:

$$\theta_2^* = \mathbb{E}[Y^{z=1,\overline{\delta}=\overline{0}}] - \mathbb{E}[Y^{z=0,\overline{\delta}=\overline{0}}].$$

Remark: As discussed in Chapter 4, if a different strategy (or a different combination of strategies) is applied to handle ICEs, then the research question in Step 1 needs to be revised accordingly. See the remark below Step 3 of Example 1 for more detail.

Step 4: Data

Conduct sample size calculation that is aligned with the causal estimand, using one of the methods discussed in Chapter 4. Then enroll patients. And then collect data following the study protocol.

Example 2 (continued). Collect data consisting of $X_i(0), Z_i, \Delta_i(t), (1 - \Delta_i(t))\widetilde{X}_i(t), (1 - \Delta_i(t))Y_i(t), t = 1, \ldots, T-1, \Delta_i(T), (1 - \Delta_i(T))Y_i$, $i = 1, \ldots, N$, where $X(0)$ includes stratification factor S, $\widetilde{X}(t)$ is a vector of time-dependent covariates, $Y(t)$ is outcome measured at t, $t = 1, \ldots, T-1$, and $Y = Y(T)$ is the outcome variable measured at T. Let $X(t)$ include both time-dependent covariates $\widetilde{X}(t)$ and intermediate outcome $Y(t)$, and let $\overline{X}(0) = X(0)$ and $\overline{X}(t) = (X(0), \ldots, X(t))$, $t = 1, \cdots, T-1$.

Step 5: Statistical estimand

Under the identifiability assumptions, translate the causal estimand into a statistical estimand.

Example 2 (continued). Under the identifiability assumptions (consistency, MAR, positivity), the causal estimand can be translated into the following statistical estimand,

$$\theta_2 = \mathbb{E}_{\overline{X} \sim \mathcal{F}_1}\left[\mathbb{E}(Y|\overline{X}, Z = 1, \overline{\Delta} = \overline{0})\right] - \mathbb{E}_{\overline{X} \sim \mathcal{F}_0}\left[\mathbb{E}(Y|\overline{X}, Z = 0, \overline{\Delta} = \overline{0})\right],$$

where the outer expectation on the right-hand-side is over $\overline{X} \sim \mathcal{F}_j$ and $\mathcal{F}_j$ is a distribution function of $(X(0), \ldots, X(T-1))$ defined as

$$\begin{aligned} &\mathbb{P}_{\mathcal{F}_j}(X(0) = x(0), \ldots, X(T-1) = x(T-1)) \\ =&\mathbb{P}\{X(0) = x(0)\} \prod_{t=1}^{T-1} \mathbb{P}\left\{X(t) = x(t)|\overline{X}(t-1) = \overline{x}(t-1), Z = j, \overline{\Delta}(t) = \overline{0}\right\}. \end{aligned}$$

Step 6: Estimator

To estimate the statistical estimand, one of the estimation methods discussed in Chapter 9 is selected.

Example 2 (continued). We will demonstrate the application of LTMLE in Example 6. In this example and Example 3, we demonstrate the application of an alternative method that combines the convenience of multiple imputation (MI) and the strength of TMLE; referred to as MI+TMLE. An advantage of using MI+TMLE is that it will be straightforward to conduct sensitivity analysis for the MAR assumption under different imputation models that reflect different MNAR scenarios.

Step 7: Estimate

Plugging the data into the proposed estimator, a point estimate is obtained, along with standard error estimator, confidence estimate, and p-value.

Example 2 (continued). According to the name, MI+TMLE consists of two stages. In stage one, missing values are imputed using some multiple imputation procedure under the MAR assumption (say, using the R package "mice"). In stage two, TMLE is applied to the imputed dataset, in which there is no need to specify arguments `Delta` or `g.Delta.SL.library`. Repeat the process by M times and combine the results using Rubin's rule.

Step 8: Sensitivity analysis

Conduct sensitivity analysis to explore the robustness of the results from the main estimator to deviations from its underlying assumptions.

Example 2 (continued). The identifiability assumptions are the consistency, MAR, and positivity assumptions. Reference-based imputation methods discussed in Chapter 10, including the copy-reference (CR) imputation and jump-to-reference (J2R) imputation, are appropriate for conducting sensitivity analysis to explore the robustness of the results under MNAR scenarios.

To implement the CR imputation, consider the following steps. First, obtain the subset consisting of all the subjects with $Z_i = 0$ and all the subjects with $Z_i = 1$ and $\Delta_i(T) = 1$, keeping the remaining set consisting of those subjects with $Z_i = 1$ and $\Delta_i(T) = 0$ for later use. Second, apply some MI procedure under the MAR assumption to the subset. Third, combine the imputed subset with the aforementioned set consisting of those subjects with $Z_i = 1$ and $\Delta_i(T) = 0$. Fourth, apply TMLE to the final full dataset.

To implement the J2R imputation, consider the following steps. First, obtain the subset consisting of all the subjects with $Z_i = 0$ and all the subjects with $Z_i = 1$ and $\Delta_i(T) = 1$, keeping the remaining set consisting of those subjects with $Z_i = 1$ and $\Delta_i(T) = 0$ for later use. Second, delete the observed outcome data in the subset for those subjects with $Z_i = 1$ and $\Delta_i(T) = 1$, bearing in mind that these tentatively deleted data will be restored. Third, apply some MI procedure under the MAR assumption to the modified subset. Fourth, restore the tentatively-deleted outcome data back to the imputed subset and combine the subset with the remained set consisting of those subjects with $Z_i = 1$ and $\Delta_i(T) = 0$. Fifth, apply TMLE to the final full dataset.

Repeat the above process, either CR or J2R, by M times and combine the results using Rubin's rule.

Step 9: Evidence

The last step is to interpret the results of the main analysis and the corresponding sensitivity analysis.

Example 2 (continued). If the results show that the treatment effect is statistically significant under MAR and statistically significant under those pre-specified MNAR scenarios, then the evidence is robust. However, if the results show that the treatment effect is not significant under those pre-specified MNAR scenarios, then the evidence is not robust.

12.2.3 RCTs with time-to-event outcome

In the preceding subsection, the outcome variable is either continuous or binary. As discussed in Chapter 5, an RCT with time-to-event outcome can be converted into a longitudinal RCT with repeatedly measured binary outcome. Following this idea, we only need to slightly revise the roadmap in the preceding subsection and apply it to RCTs with time-to-event outcome.

Step 1: Research question

Follow the PROTECT criteria to prepare the following items:

- Population: defined via inclusion/exclusion criteria;
- Response/Outcome variable: survival outcome Y;
- Treatment/Exposure variable: treatment 1 or 0;
- Counterfactual thinking: randomization and blinding;
- Time: survival status measured at $t = 1, \ldots, T$.

Research question/objective: to establish the existence, and estimate the magnitude, of treatment effects of treatment 1 vs. treatment 0 on the survival rate at time T counting from the time of treatment initiation among all the subjects in the well-defined ITT population.

Note that the population-level summary, which is the survival rate at time T in the above, can be replaced by others, say restricted mean survival time (RMST) limited by T.

Step 2: Study design

Consider a complete RCT or stratified RCT with stratification factor S. In a typical double-blind RCT, treatment 0 is placebo.

Step 3: Causal estimand

Following Chapter 5, define a causal estimand in terms of potential outcomes, taking into account the handling of the anticipated censoring and ICEs.

Example 3. In a stratified RCT, assume that there are two types of ICEs, treatment dropout and analysis dropout (also known as right censoring for time-to-event outcome), and assume that the treatment policy strategy is applied to handle treatment dropout and the hypothetical strategy—envisaging a hypothetical scenario in which censoring would not occur—is applied to handle censoring. Since the treatment policy strategy is applied to handle treatment dropout (say, treatment discontinuation, rescue medication), the "T/E" component of the research question in Step 1 needs to be revised, incorporating treatment discontinuation and rescue medication as parts of the treatments under comparison.

Let Z be the treatment assignment. Let Y^* be the time-to-death. Define $Y(t) = I(Y^* \leq t)$, with $Y(t) = 1$ indicating death status at t. Let $\overline{\Delta} =$

$(\Delta(1), \ldots, \Delta(T-1))$ be the vector of the indicators of censoring up to $T-1$. In the definition of $\overline{\Delta}$, $\Delta(T)$ is excluded because $\Delta(T) = 1$ implies that $Y(T) = 0$. Let $Y = Y(T)$ be the primary outcome, which is the death/survival status at T. By convention, if $Y(t) = 1$, then $Y(t') = 1$ and $\Delta(t') = 0$ for all $t' \geq t$. Assume we are interested in the following causal estimand:

$$\theta_3^* = \mathbb{E}[Y^{z=1,\bar{\delta}=\bar{0}}] - \mathbb{E}[Y^{z=0,\bar{\delta}=\bar{0}}].$$

An alternative version of Example 3. Same as Example 3, except that the hypothetical strategy—envisaging a hypothetical scenario in which neither treatment dropout nor censoring would occur—is applied to handle both treatment dropout and censoring. Let $\widetilde{\Delta}(t) = 1$ if either treatment dropout or censoring occurs at time t. The remaining steps will be the same except that $\Delta(t)$ is replaced by $\widetilde{\Delta}(t)$.

Step 4: Data

Conduct sample size calculation that is aligned with the causal estimand. Then enroll patients. And then collect data following the study protocol.

Example 3 (continued). Collect data consisting of $X_i(0), Z_i, \Delta_i(t), (1 - \Delta_i(t))X_i(t), (1 - \Delta_i(t))Y_i(t), t = 1, \ldots, T-1, \Delta_i(T), (1 - \Delta_i(T))Y_i$, $i = 1, \ldots, N$, where $X(0)$ includes stratification factor S, $X(t)$ is vector of time-dependent covariates, $Y(t)$ is death/survival status at t, $t = 1, \ldots, T-1$, and $Y = Y(T)$ is the death/survival status at T. By convention, if $Y(t) = 1$, then let $Y(t') = 1$, $X(t') = X(t)$, and $\Delta(t') = 0$ for all $t' \geq t$.

Step 5: Statistical estimand

Under the identifiability assumptions, translate the causal estimand into a statistical estimand.

Example 3 (continued). Under the identifiability assumptions (consistency, censoring at random (CAR), positivity), the causal estimand can be translated into the following statistical estimand,

$$\theta_3 = \mathbb{E}_{\overline{X} \sim \mathcal{F}_1}\left[\mathbb{E}(Y|\overline{X}, Z = 1, \overline{\Delta} = \overline{0})\right] - \mathbb{E}_{\overline{X} \sim \mathcal{F}_0}\left[\mathbb{E}(Y|\overline{X}, Z = 0, \overline{\Delta} = \overline{0})\right],$$

where the outer expectation on the right-hand-side is over $\overline{X} \sim \mathcal{F}_j$ and $\mathcal{F}_j$ is the same as the one defined in the preceding subsection.

Step 6: Estimator

To estimate the statistical estimand, one of the estimation methods discussed in Chapter 9 is selected.

Example 3 (continued). We will demonstrate the application of LTMLE in Example 6. Like in Example 2, in this example we consider MI+TMLE.

Step 7: Estimate

Plugging the data into the proposed estimator, a point estimate is obtained, along with standard error estimate, 95% confidence estimate, and p-value.

Example 3 (continued). Like in Example 2, MI+TMLE consists of two stages. In stage one, censored values are imputed using some multiple imputation procedure under the CAR assumption. After each imputation, we need to revise the imputed values following the convention: if $Y(t) = 1$, then let $Y(t') = 1$ and $X(t') = X(t)$ for all $t' \geq t$. In stage two, TMLE is applied to the imputed dataset. Repeat the process by M times and combine the results using Rubin's rule.

Step 8: Sensitivity analysis

Conduct sensitivity analysis to explore the robustness of the results from the main estimator to deviations from its underlying assumptions.

Example 3 (continued). The identifiability assumptions are the consistency, CAR, and positivity assumptions. Same as in Example 2, reference-based imputation methods (say, CR and J2R imputation methods) are appropriate methods for conducting sensitivity analysis to explore the robustness of the results under censoring not at random (CNAR) scenarios.

Step 9: Evidence

The last step is to interpret the results of the main analysis and the corresponding sensitivity analysis.

Example 3 (continued). If the main results show that the treatment effect is statistically significant under CAR and statistically significant under those pre-specified CNAR scenarios, then the evidence is robust. However, if the treatment effect is not statistically significant under those pre-specified CNAR scenarios, then the evidence is not robust.

12.3 Applications to Cohort Studies

Consider a sample of subjects to be enrolled in an observational study. Although the sample is usually sampled by convenience, the subjects in the sample can be assumed to be i.i.d. drawn from a population which is conceptually constructed—like the way by which a super-population is constructed. Therefore, as in Chapters 6–9, we denote the sample size as n.

The main difference between an interventional study and an observational cohort study is that the propensity score function is unknown in an observational study.

12.3.1 Cohort studies with a single follow-up

We have discussed this application in Chapter 11 in detail using a plasmode case study. Here we briefly go through the steps.

Step 1: Research question

Follow the PROTECT criteria to prepare the following items:

- Population: defined via inclusion/exclusion criteria;
- Response/Outcome variable: Y;
- Treatment/Exposure variable: treatment 1 or 0;
- Counterfactual thinking: confounding and confounders;
- Time: outcome measured at T.

Research question/objective: to establish the existence, and estimate the magnitude, of treatment effects of treatment 1 vs. treatment 0 on the outcome variable measured at time T from the time of treatment initiation among all the subjects in the population.

Step 2: Study design

Consider a cohort study, consisting of two unmatched cohorts, one cohort treated by treatment 1 and the other cohort treated by treatment 0. Baseline is the time when the treatment is initiated and follow-up time T is the time when the outcome variable is measured. The study can be prospective or retrospective, which should be determined at this step.

Step 3: Causal estimand

Following Chapter 6, define a causal estimand in terms of potential outcomes. Following Chapters 3–4, revise the causal estimand to take into account the handling of missing data and ICEs.

Example 4. Assume that in a cohort study, there are two types of ICEs, treatment discontinuation and analysis dropout, and assume that the hypothetical strategy—envisaging a hypothetical scenario in which neither ICE would occur—is applied to handle the ICEs. Let A be the treatment assignment and let Δ be the indicator of any ICE. Assume we are interested in the following causal estimand:

$$\theta_4^* = \mathbb{E}[Y^{a=1,\delta=0}] - \mathbb{E}[Y^{a=0,\delta=0}].$$

Remark: As discussed in Chapter 4, if a different strategy (or a different combination of strategies) is applied to handle ICEs, then the research question in Step 1 needs to be revised accordingly. See the remark below Step 3 of Example 1 for more detail.

Step 4: Data

If the cohort study is prospective, conduct sample size calculation that is aligned with the causal estimand as discussed in Section 10.4. Then enroll patients. And then collect data following the study protocol.

If the cohort study is retrospective, conduct feasibility analysis to determine the feasible sample size. Then obtain the data (say, from preexisting databases) following the study protocol.

Example 4 (continued). Collect or obtain data

$$\mathcal{O} = \{(X_i, A_i, \Delta_i, (1-\Delta_i)Y_i, i = 1, \ldots, n\},$$

where X consists of pre-treatment covariates that ensure the exchangeability assumption and the MAR assumption be satisfied and covariates that are believed to be associated with the outcome variable.

Step 5: Statistical estimand

Under the identifiability assumptions, translate the causal estimand into a statistical estimand.

Example 4 (continued). Under the identifiability assumptions (consistency, exchangeability, MAR, positivity), the causal estimand can be translated into the following statistical estimand,

$$\theta_4 = \mathbb{E}[Y|X, A = 1, \Delta = 0] - \mathbb{E}[Y|X, A = 0, \Delta = 0].$$

Step 6: Estimator

To estimate the statistical estimand, one of the estimation methods discussed in Chapters 7–8 is selected.

Example 4 (continued). TMLE is doubly robust; it is consistent if either the propensity score function and the non-missing probability function are estimated consistently or the outcome regression function is estimated consistently. TMLE is also an RAL estimator with the efficient influence function.

Step 7: Estimate

Plugging the data into the proposed estimator, a point estimate is obtained, along with standard error estimate, 95% confidence estimate, and p-value.

Example 4 (continued). To implement TMLE, consider the following R function "tmle."

```
tmle_fit <- tmle(Y = Y, A = A, W = W, Delta = 1 - Delta, Q.SL.
    library = c("SL.glm", "tmle.SL.dbarts2", "SL.glmnet"), g.SL.
    library = c("SL.glm", "tmle.SL.dbarts.k.5", "SL.gam"), g.Delta
    .SL.library = c("SL.glm", "tmle.SL.dbarts.k.5", "SL.gam"),
    family = "gaussian", ...)

tmle_fit$estimates$ATE$psi
tmle_fit$estimates$ATE$CI
```

The above R codes are the same as those in Example 1, except that `g.SL.library` is used to replace `g1W` to specify a library of predictive models to fit super learner for the propensity score function, which is unknown in the observational cohort study.

Step 8: Sensitivity analysis

Conduct sensitivity analysis to explore the robustness of the results from the main estimator to deviations from its underlying assumptions.

Example 4 (continued). The identifiability assumptions are the consistency, exchangeability, MAR, and positivity assumptions. Among them, the exchangeability assumption (a.k.a., the no-unmeasured confounding assumption) is the most crucial one for non-randomized studies. For example, E-value is an appropriate method to explore the robustness of the results if there are unmeasured confounders.

Step 9: Evidence

The last step is to interpret the results of the main analysis and the corresponding sensitivity analysis. The findings from such interpretation are summarized as evidence.

Example 4 (continued). In order to interpret the E-value, some threshold (say, 2) should be pre-specified in the protocol that is agreed upon among the stakeholders. If the E-value is larger than or equal to the pre-specified threshold, the main result can be claimed to be robust. Otherwise, the main result is not robust.

12.3.2 Externally controlled trials

An externally controlled trials (ECTs) can be considered as a special point-exposure cohort study, which consists of two cohorts—the treated cohort is the single-arm clinical trial and the control cohort is formed externally. The control cohort may be formed based on the placebo arms of some historical RCTs or may be formed based on external real-world data (RWD).

But there are some differences between the ECT to be discussed soon in the following and the typical cohort study that is discussed in the preceding example. In the preceding cohort study, the population consists of all the subjects to be treated by either the investigative treatment or the control treatment, and therefore the estimand is the average treatment effect (ATE). In the ECT to be discussed soon, the population consists of all the subjects that are treated by the investigative treatment in the real world, and therefore the estimand is the average treatment effect among the treated (ATT).

Step 1: Research question

Follow the PROTECT criteria to prepare the following items:

- Population: defined for the single-arm clinical trial;
- Response/Outcome variable: Y;
- Treatment/Exposure variable: treatment 1 or 0;
- Counterfactual thinking: confounding and confounders;
- Time: outcome measured at T.

Research question/objective: to establish the existence, and estimate the magnitude, of treatment effects of treatment 1 vs. treatment 0 on the outcome variable measured at time T from the time of treatment initiation among the subjects who are treated by treatment 1 in the real world.

Step 2: Study design

Consider an ECT, consisting of a single-arm clinical trial and an external control arm. The subjects of the external control arm may be from the placebo arms of some historical RCTs or may be from RWD.

Step 3: Causal estimand

Following Chapter 6, define a causal estimand in terms of potential outcomes. Following Chapters 3–4, revise the causal estimand to take into account the handling of missing data and ICEs.

Example 5. Assume that in a cohort study, there are two types of ICEs, treatment discontinuation and analysis dropout, and assume that the hypothetical strategy—envisaging a hypothetical scenario in which neither ICE would occur—is applied to handle the ICEs. Let A be the treatment assignment and let Δ be the indicator of any ICE. Assume we are interested in the following causal estimand:

$$\theta_5^* = \mathbb{E}[Y^{a=1,\delta=0}|A=1] - \mathbb{E}[Y^{a=0,\delta=0}|A=1].$$

Step 4: Data

Since the single-arm clinical trial is prospective, conduct sample size calculation that is aligned with the causal estimand as discussed in Section 4.4 and Section 10.4. Then enroll patients. And then collect data following the study protocol.

Meanwhile, since the data that form the external control arm are from retrospective data sources, conduct feasibility analysis to determine the feasible sample size. Then obtain the data (say, from preexisting databases) following the study protocol.

Example 5 (continued). Collect or obtain data

$$\mathcal{O} = \{(X_i, A_i, \Delta_i, (1-\Delta_i)Y_i, i = 1, \ldots, n\},$$

where A is a binary variable with $A = 1$ indicating the subject is from the single-arm clinical trial and $A = 0$ indicating the subject is from the external control arm, X consists of pre-treatment covariates that ensure the exchangeability assumption and the MAR assumption be satisfied and covariates that are believed to be associated with the outcome variable.

Step 5: Statistical estimand

Under the identifiability assumptions, translate the causal estimand into a statistical estimand.

Example 5 (continued). Under the identifiability assumptions (consistency, exchangeability, MAR, positivity), the causal estimand can be translated into the following statistical estimand,

$$\theta_5 = \mathbb{E}_{X|A=1}[Y|X, A = 1, \Delta = 0] - \mathbb{E}_{X|A=1}[Y|X, A = 0, \Delta = 0],$$

where the subscript $X|A = 1$ means that the expectation is taken over the distribution of X among subjects in the single-arm clinical trials.

Step 6: Estimator

To estimate the statistical estimand, one of the estimation methods discussed in Chapters 7–8 is selected.

Example 5 (continued). Same as in Example 4, TMLE is selected.

Step 7: Estimate

Plugging the data into the proposed estimator, a point estimate is obtained, along with standard error estimate, 95% confidence estimate, and p-value.

Example 5 (continued). To implement TMLE, consider the following R function "tmle."

```
tmle_fit <- tmle(Y = Y, A = A, W = W, Delta = 1 - Delta, Q.SL.library = c("SL.glm", "tmle.SL.dbarts2", "SL.glmnet"), g.SL.library = c("SL.glm", "tmle.SL.dbarts.k.5", "SL.gam"), g.Delta.SL.library = c("SL.glm", "tmle.SL.dbarts.k.5", "SL.gam"), family = "gaussian", ...)

tmle_fit$estimates$ATT$psi
tmle_fit$estimates$ATT$CI
```

The above R codes are the same as those in Example 4, except that the ATT estimates instead of the ATE estimates are reported.

Step 8: Sensitivity analysis

Conduct sensitivity analysis to explore the robustness of the results from the main estimator to deviations from its underlying assumptions.

Example 5 (continued). Same as Example 4.

Step 9: Evidence

The last step is to interpret the results of the main analysis and the corresponding sensitivity analysis. The findings from such interpretation are summarized as the evidence.

Example 5 (continued). Same as Example 4.

12.3.3 Longitudinal cohort studies

Step 1: Research question

Follow the PROTECT criteria to prepare the following items:

- Population: defined via inclusion/exclusion criteria;
- Response/Outcome variable: Y;
- Treatment/Exposure variable: treatment regimes $\overline{1}$ or $\overline{0}$;
- Counterfactual thinking: time-dependent confounding;
- Time: outcome measured at $t = 1, \ldots, T$.

Research question/objective: to establish the existence, and estimate the magnitude, of treatment effects of treatment regime $\overline{a} = \overline{1}$ vs. $\overline{a} = \overline{0}$ among all the subjects in the population.

Note that in the above research question, the comparison between $\overline{a} = \overline{1}$ vs. $\overline{a} = \overline{0}$ can be replaced by the comparison between any other two feasible treatment regimes.

Also note that the primary outcome variable can be continuous, binary, or time-to-event, bearing in mind that a time-to-event outcome can be converted into a longitudinally measured binary outcome.

Step 2: Study design

Consider a longitudinal cohort study, consisting of two unmatched cohorts, one cohort treated by treatment 1 initially and the other cohort treated by treatment 0 initially. Baseline is the time when the treatment is initiated and there are T follow-up visits.

The study can be prospective or retrospective or a hybrid of prospective and retrospective, which should be determined at this step.

Step 3: Causal estimand

Following Chapter 6, define a causal estimand in terms of potential outcomes, taking into account the handling of missing data and ICEs.

Example 6. In a longitudinal cohort study, the hypothetical strategy is applied to handle treatment dropouts and analysis dropouts. Let $\overline{A} = (A(0), \ldots, A(T-1))$ be the treatment sequence and let $\overline{\Delta} = (\Delta(1), \ldots, \Delta(T))$ be the vector of the indicators of ICE occurrences. Assume we are interested in the following causal estimand:

$$\theta_6^* = \mathbb{E}[Y^{\overline{a}=\overline{1},\overline{\delta}=\overline{0}}] - \mathbb{E}[Y^{\overline{a}=\overline{0},\overline{\delta}=\overline{0}}].$$

Step 4: Data

Conduct sample size calculation that is aligned with the causal estimand. Then enroll patients. And then collect data following the study protocol.

Example 6 (continued). Collect data consisting of $X_i(0), A_i(0), \Delta_i(t), (1 - \Delta_i(t))\widetilde{X}_i(t), (1 - \Delta_i(t))Y_i(t), (1 - \Delta_i(t))A_i(t), t = 1, \ldots, T-1, \Delta_i(T), (1 - \Delta_i(T))Y_i, i = 1, \ldots, n$, where $X(0)$ includes baseline covariates, $\widetilde{X}(t)$ is vector of time-dependent covariates, $Y(t)$ is the outcome measured at t, $t = 1, \ldots, T-1$, and $Y = Y(T)$ is the outcome variable measured at T. Let $X(t)$ include both time-dependent covariates $\widetilde{X}(t)$ and intermediate outcome $Y(t)$, and let $\overline{X}(t) = X(0)$ and $\overline{X}(t) = (X(0), \ldots, X(t))$, $t = 1, \ldots, T-1$.

Step 5: Statistical estimand

Under the identifiability assumptions, translate the causal estimand into a statistical estimand.

Example 6 (continued). Under the identifiability assumptions (consistency, sequential exchangeability, MAR/CAR, positivity), the causal estimand can be translated into the following statistical estimand,

$$\theta_6 = \mathbb{E}_{\overline{X} \sim \mathcal{F}_{\overline{1}}}\left[\mathbb{E}(Y|\overline{X}, \overline{A} = \overline{1}, \overline{\Delta} = \overline{0})\right] - \mathbb{E}_{\overline{X} \sim \mathcal{F}_{\overline{0}}}\left[\mathbb{E}(Y|\overline{X}, \overline{A} = \overline{0}, \overline{\Delta} = \overline{0})\right],$$

where the outer expectation on the right-hand-side is over $\overline{X} \sim \mathcal{F}_{\overline{a}}$ and where $\mathcal{F}_{\overline{a}}$ is the following distribution function of $\overline{X}$,

$$\begin{aligned} &\mathbb{P}_{\mathcal{F}_{\overline{a}}}(X(0) = x(0), \ldots, X(T-1) = x(T-1)) \\ =&\mathbb{P}\{X(0) = x(0)\} \prod_{t=1}^{T-1} \mathbb{P}\left\{X(t) = x(t)|\overline{X}(t-1) = \overline{x}(t-1), \overline{A}(t) = \overline{a}, \overline{\Delta}(t) = \overline{0}\right\}. \end{aligned}$$

Step 6: Estimator

To estimate the statistical estimand, one of the estimation methods discussed in Chapter 9 is selected.

Example 6 (continued). To estimate θ_6, LTMLE is an efficient estimator.

Step 7: Estimate

Plugging the data into the proposed estimator, obtain the point estimate, along with standard error estimate, confidence interval estimate, and p-value.

Example 6 (continued). To implement LTMLE, consider the following R function "ltmle."

```
# nn is sample size
# nt is num of decision points
treatment <- matrix(1, nrow = nn, ncol = nt)
control <- matrix(0, nrow == nn, ncol = nt)

result <- ltmle(data, Anodes, Cnodes, Lnodes, Ynodes,
```

```
survivalOutcome = NULL, Yrange = NULL,
Qform = NULL, gform = NULL, SL.library = "glm",
abar = list(treatment, control), rule = NULL, ...)

print(summary(result))
```

In the above R codes, two treatment regimes under comparison are defined as `treatment` and `control`. In "ltmle," argument `Anodes` is to specify column names corresponding to treatment nodes $\overline{A}$, with argument `Cnodes` for column names corresponding to missing/censoring nodes $\overline{\Delta}$, argument `Lnodes` for column names corresponding to time-dependent covariates $\overline{X}$, and `Ynodes` for column names corresponding to outcome nodes $Y(1), \ldots, Y(T)$. Argument `abar` is to specify the comparison pair of treatment regimes.

If all the values of outcome nodes are 0 or 1, then the outcome is binary. If some values of outcome nodes are not 0 or 1 and `survivalOutcome==FALSE`, then the outcome variable is continuous. If the outcome variable is continuous and `Yrange` is specified, then the outcome variable is transformed to $(Y - \min(\text{Yrange}))/(\max(\text{Yrange}) - \min(\text{Yrange}))$ and at the end, parameter estimates are transformed back based on `Yrange`. If `survivalOutcome==TRUE` is specified, then the outcome is time-to-event.

Argument `Qform` is to specify regression formulas for the outcome regression functions Q's and argument `gform` is to specify regression formulas for the propensity score functions g's and non-missing probability functions h's. If `Qform=NULL` and `gform=NULL`, then use `SL.library` to specify a list of libraries to fit super learners to Q's, g's, and h's.

Step 8: Sensitivity analysis

Conduct sensitivity analysis to explore the robustness of the results from the main estimator to deviations from its underlying assumptions.

Example 6 (continued). The identifiability assumptions are the consistency, sequential exchangeability, MAR/CAR, and positivity assumptions. For example, E-value is an appropriate method to explore the robustness of the results under the assumption that there are unmeasured confounders and reference-based imputation methods can be used to explore the robustness of the results under some pre-specified MNAR/CNAR scenarios.

Step 9: Evidence

The last step is to interpret the results of the main analysis and the corresponding sensitivity analysis.

Example 6 (continued). In order to interpret the E-value, some threshold (say, 2) should be pre-specified in the protocol that is agreed upon among the stakeholders. If the E-value is larger than or equal to the pre-specified threshold and the results from those pre-specified MNAR/CNAR scenarios remain significant, then the main results can be claimed to be robust. Otherwise, the main results are not robust.

12.4 Exercises

Ex 12.1

Explain that LTMLE is an appropriate method to replace MI+TMLE in Step 6 of Example 2. Write some key arguments for the R function "ltmle" to implement LTMLE in Step 7 of Example 2.

Ex 12.2

Explain that LTMLE is an appropriate method to replace MI+TMLE in Step 6 of Example 3. Write some key arguments for the R function "ltmle" to implement LTMLE in Step 7 of Example 3.

Bibliography

Armitage, Peter. 2003. "Fisher, Bradford Hill, and randomization." *International Journal of Epidemiology* 32 (6): 925–928.

Austin, Peter, Douglas Lee, and Jason Fine. 2016. "Introduction to the analysis of survival data in the presence of competing risks." *Circulation* 133 (6): 601–609.

Benkeser, David, Marco Carone, and Peter Gilbert. 2018. "Improved estimation of the cumulative incidence of rare outcomes." *Statistics in Medicine* 37 (2): 280–293.

Bickel, Peter, Chris Klaassen, Ya'acov Ritov, and Jon Wellner. 1993. *Efficient and Adaptive Estimation for Semiparametric Models.* Springer.

Brookhart, Alan, Sebastian Schneeweiss, Kenneth Rothman, Robert Glynn, Jerry Avorn, and Til Stürmer. 2006. "Variable selection for propensity score models." *American Journal of Epidemiology* 163 (12): 1149–1156.

Chow, Shein-Chung, Jun Shao, Hansheng Wang, and Yuliya Lokhnygina. 2017. *Sample Size Calculations in Clinical Research.* CRC Press.

Chung, Kai-Lai. 2001. *A Course in Probability Theory.* Academic Press.

Dawid, Philip. 2010. "Beware of the DAG!" In *Causality: Objectives and Assessment,* 59–86. PMLR.

Deming, Edwards, and Frederick Stephan. 1941. "On the interpretation of censuses as samples." *Journal of the American Statistical Association* 36 (213): 45–49.

Ding, Peng, and Jiannan Lu. 2017. "Principal stratification analysis using principal scores." *Journal of the Royal Statistical Society. Series B (Statistical Methodology)* 79 (3): 757–777.

Ding, Peng, and Tyler VanderWeele. 2016. "Sensitivity analysis without assumptions." *Epidemiology* 27 (3): 368.

Efron, Bradley, and Robert Tibshirani. 1994. *An Introduction to the Bootstrap.* CRC Press.

Fang, Yixin. 2020. "Two basic statistical strategies of conducting causal inference in real-world studies." *Contemporary Clinical Trials* 99: 106193.

Fang, Yixin, and Weili He. 2023a. "Key considerations in forming research questions and conducting research in real-world setting." In *Real-World Evidence in Medical Product Development,* 29–41. Springer.

———. 2023b. "Sensitivity analysis in the analysis of real-world data." In *Real-World Evidence in Medical Product Development,* 271–287. Springer.

Fang, Yixin, Weili He, Xiaofei Hu, and Hongwei Wang. 2021. "A method for sample size calculation via E-value in the planning of observational studies." *Pharmaceutical Statistics* 20 (1): 163–174.

Fang, Yixin, Weili He, Hongwei Wang, and Meijing Wu. 2020. "Key considerations in the design of real-world studies." *Contemporary Clinical Trials* 96: 106091.

Fang, Yixin, and Man Jin. 2021. "Sample size calculation when planning clinical trials with intercurrent events." *Therapeutic Innovation & Regulatory Science* 55: 779–785.

———. 2022. "Sequential modeling for a class of reference-based imputation methods in clinical trials with quantitative or binary outcomes." *Statistics in Medicine* 41 (8): 1525–1540.

Fang, Yixin, Man Jin, and Chengqing Wu. (2024). "Aligning sample size calculations with estimands in clinical trials with time-to-event outcomes." *Statistics and Its Interface* 17 (1): 63–68.

Fang, Yixin, Hongwei Wang, and Weili He. 2020. "A statistical roadmap for journey from real-world data to real-world evidence." *Therapeutic Innovation & Regulatory Science* 54: 749–757.

Frangakis, Donald, Constantineand Rubin. 2002. "Principal stratification in causal inference." *Biometrics* 58 (1): 21–29.

Gaarder, Jostein. 1994. *Sophie's World: A Novel about the History of Philosophy.* Macmillan.

Greenland, Sander, Judea Pearl, and James Robins. 1999. "Causal diagrams for epidemiologic research." *Epidemiology* 10 (1): 37–48.

Gruber, Susan, and Mark van der van der Laan. 2012. "tmle: An R package for targeted maximum likelihood estimation." *Journal of Statistical Software* 51: 1–35.

Hastie, Trevor, Robert Tibshirani, and Jerome Friedman. 2009. *The Elements of Statistical Learning: Data Mining, Inference, and Prediction.* Vol. 2. Springer.

Hernán, Miguel, and James Robins. 2020. *Causal Inference: What If.* Boca Raton: Chapman & Hall/CRC.

Huber, Peter. 1964. "Robust estimation of a location parameter." *Annals of Mathematical Statistics* 35: 73–101.

Imbens, Guido, and Donald Rubin. 2015. *Causal Inference in Statistics, Social, and Biomedical Sciences.* Cambridge University Press.

Jiang, Zhichao, Shu Yang, and Peng Ding. 2022. "Multiply robust estimation of causal effects under principal ignorability." *Journal of the Royal Statistical Society Series B: Statistical Methodology* 84 (4): 1423–1445.

Kalbfleisch, John, and Ross Prentice. 2011. *The Statistical Analysis of Failure Time Data.* John Wiley & Sons.

Kummerfeld, Erich, Jaewon Lim, and Xu Shi. 2022. "Data-driven automated negative control estimation (DANCE): Search for, validation of, and causal inference with negative controls." *arXiv preprint arXiv:2210.00528.*

Lendle, Samuel, Joshua Schwab, Maya Petersen, and Mark van der Laan. 2017. "ltmle: An R package implementing targeted minimum loss-based estimation for longitudinal data." *Journal of Statistical Software* 81: 1–21.

Linden, Ariel, Maya Mathur, and Tyler VanderWeele. 2020. "Conducting sensitivity analysis for unmeasured confounding in observational studies using E-values: The evalue package." *The Stata Journal* 20 (1): 162–175.

Mallinckrodt, Craig, Geert Molenberghs, Ilya Lipkovich, and Bohdana Ratitch. 2020. *Estimands, Estimators and Sensitivity Analysis in Clinical Trials.* CRC Press.

Miao, Wang, Zhi Geng, and Eric Tchetgen Tchetgen. 2018. "Identifying causal effects with proxy variables of an unmeasured confounder." *Biometrika* 105 (4): 987–993.

Murphy, Susan, Mark van der Laan, James Robins, and Conduct Problems Prevention Research Group. 2001. "Marginal mean models for dynamic regimes." *Journal of the American Statistical Association* 96 (456): 1410–1423.

National Research Council. 2010. *The Prevention and Treatment of Missing Data in Clinical Trials.* National Academies Press.

Newey, Whitney. 1990. "Semiparametric efficiency bounds." *Journal of Applied Econometrics* 5 (2): 99–135.

———. 1994. "The asymptotic variance of semiparametric estimators." *Econometrica* 62 (6): 1349–1382.

Neyman, Jerzy. 1923. "On the application of probability theory to agricultural experiments. Essay on principles." *Annals Agricultural Sciences,* 1–51.

O'Kelly, Michael, and Bohdana Ratitch. 2014. *Clinical Trials with Missing Data: A Guide for Practitioners.* John Wiley & Sons.

Pearl, Judea. 2009. *Causality.* Cambridge University Press.

Pearl, Judea, Madelyn Glymour, and Nicholas Jewell. 2016. *Causal Inference in Statistics: A primer.* John Wiley & Sons.

Pearl, Judea, and Dana Mackenzie. 2018. *The Book of Why: The New Science of Cause and Effect.* Basic Books.

Prentice, Ross, John Kalbfleisch, Arthur Peterson, Nancy Flournoy, Vernon Farewell, and Norman Breslow. 1978. "The analysis of failure times in the presence of competing risks." *Biometrics* 34 (4): 541–554.

Ratitch, Bohdana, and Michael O'Kelly. 2011. "Implementation of pattern-mixture models using standard SAS/STAT procedures." *Proceedings of Pharmaceutical Industry SAS User Group.*

Robins, James. 1986. "A new approach to causal inference in mortality studies with a sustained exposure period—application to control of the healthy worker survivor effect." *Mathematical Modelling* 7 (9-12): 1393–1512.

Rosenbaum, Paul, and Donald Rubin. 1983. "The central role of the propensity score in observational studies for causal effects." *Biometrika* 70 (1): 41–55.

Royston, Patrick, and Mahesh Parmar. 2011. "The use of restricted mean survival time to estimate the treatment effect in randomized clinical trials when the proportional hazards assumption is in doubt." *Statistics in Medicine* 30 (19): 2409–2421.

Rubin, Donald. 1975. "Bayesian inference for causality: The importance of randomization." In *The Proceedings of the social statistics section of the American Statistical Association,* 233:239. American Statistical Association Alexandria, VA.

———. 1976. "Inference and missing data." *Biometrika* 63 (3): 581–592.

———. 1980. "Discussion on "Randomization analysis of experimental data: The Fisher randomization test" by Basu." *Journal of the American statistical association* 75 (371): 591–593.

———. 1996. "Multiple imputation after 18+ years." *Journal of the American statistical Association* 91 (434): 473–489.

———. 2004. *Multiple Imputation for Nonresponse in Surveys.* John Wiley & Sons.

Schneeweiss, Sebastian. 2006. "Sensitivity analysis and external adjustment for unmeasured confounders in epidemiologic database studies of therapeutics." *Pharmacoepidemiology and Drug Safety* 15 (5): 291–303.

Stein, Charles. 1956. "Efficient nonparametric testing and estimation." In *Proceedings of the Third Berkeley Symposium on Mathematical Statistics and Probability, Volume 1: Contributions to the Theory of Statistics,* 3: 187–196. University of California Press.

Stitelman, Ori, Victor De Gruttola, and Mark van der Laan. 2012. "A general implementation of TMLE for longitudinal data applied to causal inference in survival analysis." *The International Journal of Biostatistics* 8 (1).

Tchetgen Tchetgen, Eric, Andrew Ying, Yifan Cui, Xu Shi, and Wang Miao. 2020. "An introduction to proximal causal learning." *arXiv preprint arXiv:2009.10982.*

Thorpe, Kevin, Merrick Zwarenstein, Andrew Oxman, Shaun Treweek, Curt Furberg, Douglas Altman, Sean Tunis, Eduardo Bergel, Ian Harvey, David Magid, et al. 2009. "A pragmatic–explanatory continuum indicator summary (PRECIS): a tool to help trial designers." *Journal of Clinical Epidemiology* 62 (5): 464–475.

Tsiatis, Anastasios. 2006. *Semiparametric Theory and Missing Data.* Springer.

Tsiatis, Anastasios, Marie Davidian, Shannon Holloway, and Eric Laber. 2020. *Dynamic Treatment Regimes: Statistical Methods for Precision Medicine.* Chapman & Hall/CRC.

van der Laan, Mark, Eric Polley, and Alan Hubbard. 2007. "Super learner." *Statistical Applications in Genetics and Molecular Biology* 6 (1).

van der Laan, Mark, and Sherri Rose. 2011. *Targeted Learning: Causal Inference for Observational and Experimental Data.* Springer Science & Business Media.

van der Vaart, Aad. 2000. *Asymptotic Statistics.* Cambridge University Press.

van der Laan, Mark, and Daniel Rubin. 2006. "Targeted maximum likelihood learning." *The International Journal of Biostatistics* 2 (1).

VanderWeele, Tyler, and Peng Ding. 2017. "Sensitivity analysis in observational research: introducing the E-value." *Annals of Internal Medicine* 167 (4): 268–274.

VanderWeele, Tyler, Peng Ding, and Maya Mathur. 2019. "Technical considerations in the use of the E-value." *Journal of Causal Inference* 7 (2): 20180007.

Zhang, Xiang, Douglas Faries, Hu Li, James Stamey, and Guido Imbens. 2018. "Addressing unmeasured confounding in comparative observational research." *Pharmacoepidemiology and Drug Safety* 27 (4): 373–382.

Zhang, Xiang, James Stamey, and Maya Mathur. 2020. "Assessing the impact of unmeasured confounders for credible and reliable real-world evidence." *Pharmacoepidemiology and Drug Safety* 29 (10): 1219–1227.

Index

AIPW, 120
AIPW-SL, 145
analysis dropout, 40
asymptotic, 15
    linearity, 16, 122
    normality, 15
    variance, 120
average treatment effect, 103
    among the treated, 113
    ATC, 115
    ATE, 103
    ATT, 113
    on the controls, 115

bias, 102
blinding, 19
block randomization, 33
bootstrap, 127

case-control, 100
causal estinand, 21
censoring, 72
central limit theorem, 12
central question, 1
chain, 13
claims data, 101
clever-covariate, 156
cohort, 100
collider, 14
complete randomization, 22
composite variable strategy, 59
confidence interval, 16
confounder, 102
confounding Bias, 101
consistency assumption, 4, 22, 182
consistent estimator, 15
copy reference, 186
counterfactual outcome, 2
covariates, 5
Cramer-Rao bound, 122
cross-sectional, 100
cross-validation, 144
CSR, 192

data collection, 194
decision rule, 94
direct effect, 20
directed acyclic graph (DAG), 13
double robustness, 139
double-blind, 20
doubly robust estimator, 120
duality, 9
dynamic treatment regime, 93

E-value, 183
efficient, 140
    estimator, 140
    influence function, 140
electronic health records, 101
empirical distribution, 8
endogenous variable, 51
endpoint, 3
estimand, 2, 21
estimating equation, 120
estimator, 24
evidence, 195
exchangeability assumption, 103, 183
exogenous variable, 51
external control, 101
externally controlled trial (ECT), 113

Fisher information, 122
Fisher's approach, 22
fork, 14

g-computation, 84
g-formula, 84
generalized linear model, 48

historical control, 101
hypothesis, 16
hypothetical strategy, 57

identifiability, 44
  assumptions, 103
identification, 30, 45, 103
independence, 11
indirect effect, 20
individual treatment effect, 6
influence function, 122
initial estimator, 120
intent-to-treat (ITT), 8
intent-to-treat Effect, 41
intercurrent event (ICE), 48
invariance property, 125
IPW, 120
IPW-SL, 145
ITT effect, 41

jump to reference, 186

law of iterated expectations, 11
law of large numbers, 12
library, 143
likelihood, 122
log-likelihood, 122
longitudinal, 79
  cohort, 109
  study, 79
LTMLE, 167

M-estimation, 120
M-estimator, 120
machine learning, 143
MAR assumption, 43, 186
missing, 40
  at random, 42
  completely at random, 42
  data, 40
  not at random, 42
  value, 43

missing data, 50
  rate, 74
misspecification, 143
MLE, 120
MLE-SL, 145
monotonicity assumption, 65
Multiple imputation (MI), 186
multiple imputation (MI), 86

Neyman's approach, 22
NHEFS study, 195
nonparametric, 143
NUC assumption, 103

observational studies, 100
open-label, 20
optimal estimator, 137
outcome, 1
outcome regression, 119
  function, 142

parametric, 122, 143
  submodel, 145
per-protocol effect, 45
personalized medicine, 94
placebo, 2
plug-in estimator, 127
population, 3
population-level summary, 7
positivity assumption, 43, 185
potential outcome, 2, 21
potential world, 3
PP effect, 46
pragmatic, 99
precision medicine, 94
principal ignorability assumption, 65
principal stratum strategy, 63
propensity score, 119
  function, 142
PROTECT, 6
proximal, 109
proxy, 107

RAL estimator, 148
randomization, 19

randomized controlled clinical trial (RCT), 20
real world, 3
real-world data (RWD), 99
real-world evidence (RWE), 99
reference-based imputation, 186
registries, 101
regular and asymptotically linear (RAL), 126
regularity, 121, 147
RMST, 7
roadmap, 191
robustness, 42
Rubin's rule, 86

sample size calculation, 74
sandwich formula, 121
SAP, 192
score, 122
semiparametric, 143
sensitivity analysis, 180
shortcut, 157
simple randomization, 30
single-arm, 101
SMART design, 95
standard error, 29
standard of care, 2
standardization strategy, 31, 104
static treatment regime, 93
statistical estimand, 22
statistical inference, 29
stratification, 34
  factor, 34
stratified randomization, 34
structural causal model (SCM), 51
study protocol, 40, 192
super learner, 143
super-efficiency, 125
super-population, 8
supplementary analysis, 181
survival rate, 7
SUTVA, 3
systematic bias, 19

targeted learning, 154
Taylor expansion, 121
time-dependent, 79
time-to-event, 7, 72
TMLE, 120, 155
treatment, 3
  assignment, 19
  dropout, 40
  effect, 1
treatment policy strategy, 55

unmeasured confounder, 105

validity, 9
value, 167

weighting strategy, 31, 104
while on treatment strategy, 62

Z-statistic, 15

Printed in the United States
by Baker & Taylor Publisher Services